future
food

alternate
protein
for
the year
2000

2.

future food

alternate protein for the year 2000

BARBARA FORD

WILLIAM MORROW AND COMPANY, INC.
NEW YORK   1978

Copyright © 1978 by Barbara Ford

Grateful acknowledgment is made to Autumn Press for permission to reprint from *The Book of Tofu,* copyright © 1976
by William Shurtleff and Akiko Aoyagi.

Library of Congress Cataloging in Publication Data

Ford, Barbara.
    Future food.

    Bibliography: p.
    Includes index.
    1. Proteins in human nutrition.   2. Food supply.
I. Title.
TX553.P7F67   641.1'2   77-18808
ISBN 0-688-03299-0
ISBN 0-688-08299-8 pbk.

BOOK DESIGN   CARL WEISS

Printed in the United States of America.

First Edition

1   2   3   4   5   6   7   8   9   10

TO ARIELLE,

WHO THOUGHT OF IT,

AND TO DOUG,

WHO ATE IT

ACKNOWLEDGMENTS

MANY PEOPLE PROVIDED MATERIAL FOR THIS BOOK; MY thanks to all.

The help of the following was particularly valuable: Dr. Richard L. Cooper, U.S. Regional Soybean Laboratory; Dr. Nicholas D. Pintauro, Rutgers University; Roger Eisenhauer, Dr. Walter J. Wolf, D. D. Christianson, Dr. C. W. Hesseltine, Dr. Hwa L. Wang, Dr. C. W. Blessin, all of the NRRL; Howard W. Mattson, Institute of Food Technologists; Dr. Doris H. Calloway, University of California; Dr. Louis B. Rockland, WRRL; Dr. M. Wayne Adams, Michigan State University; Dr. Carl D. Clayberg, Kansas State University; Dr. John D. Axtell, Purdue University; Dr. Virgil A. Johnson, ARS; Prof. Paul J. Mattern, University of Nebraska; Prof. Mark A. Stahmann, University of Wisconsin; Prof. J. H. Mitchell, Clemson University; Gloria Burke, Leah C. Berardi, and Dr. John P. Cherry, all of the SRRL; Dr. Hugh D. Wilson, University of Wyoming; Prof. Theodore Hymowitz and Dr. Christine Newell, both of the University of Illinois; Prof. Charles W. Weber, University of Arizona; Dr. Lawrence C. Curtis; Prof. Vernon R. Young, MIT; Dr. John H. Litchfield, Battelle Columbus Laboratories; Dr. Clayton D. Callihan, Louisiana State University; Bruce Butterfield, American Meat Institute; Dr. Glenn W. Burton, ARS; Dr.

Charles H. Black, CAST; Dr. Michael H. Robinson, Smithsonian Institution; Prof. Edwin W. King, Clemson University; Jerome La Tour, North American Bait Farms; John P. Wise, NOAA; Robert J. Learson, Northeast Fisheries Center; Dr. Paul A. Sandifer, South Carolina Marine Resources Center; Virginia H. Holsinger, ERRL; Dr. John H. Weisburger, American Health Foundation; Anne S. Moffat, Prof. Donald B. Zilversmit, and Prof. T. Colin Campbell, all of Cornell University; and Prof. Lawrence Kaplan, University of Massachusetts.

contents

1

WHAT PROTEIN SHORTAGE?

> "And let them gather all the food of these good years that are coming, and lay up grain under the authority of Pharaoh for food in the cities, and let them keep it. That food shall be a reserve for the land against the seven years of famine which are to befall the land of Egypt."
>
> —GENESIS

THE SMALL TABLE IN THE MODEST SUBURBAN NEW JERSEY restaurant where my husband and I were having dinner was already covered with food, even though the entree hadn't yet arrived. A plate of cheese and a loaf of bread in the middle, a dish of olives, radishes, and celery to one side and, in front of us, individual salads. "But why do we need a book on unconventional sources of protein?" my husband asked, cutting off a generous slice of cheese and putting it on a slice of bread. I launched into the subject of the world food shortage as I finished my salad. "Oh, here come our dinners." The waitress managed to find room for the entree only by removing the salad plates and an unwanted ashtray.

"Veal?"

"Here."

She put the veal in front of my husband, fillet of sole in

front of me. The fish was so big it almost covered the dinner plate. Our conversation turned to the excellence and abundance of the dinner. After dinner, I wrapped up about one-third of the sole in a paper napkin and brought it out to our well-fed dog in the car. As we drove home, the absurdity of the restaurant scene struck me. I had been trying to explain the world food shortage in the context of a meal in a typical U.S. restaurant, where proteins and calories are in such abundance that our dog was at that moment devouring enough fish to supply much of the daily needs of an adult human. Food shortage? If anything, much of the United States seems to be experiencing a food glut. The most expensive and best-patronized restaurant in our area (we've never been there) is famed for the generous portions of its steak and lobster. An advertisement in our local paper solicits customers for New Jersey-reared "grain-fed beef" for the home freezer. Ice cream, cheese, and bakery shops are in every shopping mall.

Oh, local shortages may develop—as I write this, the Northeast is experiencing a clam shortage, with prices for available clams double those of a few years ago—but food is abundant, even if higher priced.

Most Americans, nutritionists tell us, eat enough food. In fact, they assert, many of us eat too much food, particularly protein. Americans eat 99 grams (about three and one-half ounces) of protein per day, almost twice as much as the amount of protein suggested by the United Nations Food and Agricultural Organization (FAO) and the World Health Organization (WHO) for the average adult male. The FAO/WHO put the desirable standard at 56 grams of protein for an adult male, 46 grams for an adult female. About two-thirds of the 99 grams of protein we eat is of animal origin. Our consumption of meat has been rising for the last four decades. We ate 192 pounds of meat per person in 1976, 55 pounds per person in 1940. Per capita poultry consumption zoomed from 14 pounds to 40 pounds in the same period.

Some other countries also eat an overabundance of protein, including most of the nations of Western Europe, Australia, and New Zealand. In a few of those countries, people eat more meat than we do. Many other countries have an adequate supply of protein.

The food shortage I was trying to explain as we ate our generous veal and fish dinners is principally a problem of what are known, today, as "developing countries" (a polite phrase that has replaced the former "undeveloped countries"). Most of them are in Asia and Africa. At frequent intervals, some of these nations experience severe shortages of both protein and calorie sources, and death rates mount. Even in good times, malnutrition is a chronic condition among a large segment of the population in developing countries. This past year, 1977, was one of those good periods. The world experienced its second year of bumper harvests and food stocks were high for the first time since the early 1970's. India, one of the poorest developing countries, amassed a 20-million-ton reserve of grain, the chief source of both protein and calories in that nation. Few people died of hunger in 1977. Between 1971 and 1975, however, the death rate climbed by two million, an increase attributed to starvation or diseases associated with severe malnutrition by the Worldwatch Institute, a private, nonprofit research institution with offices in Washington, D.C. Most of those who died were children. And for every person who died, scores or even hundreds suffered various degrees of malnutrition.

Another research group, the National Research Council, which is affiliated with the National Academy of Sciences, estimates the current number of malnourished people in the world today, during a period of bumper harvests, at between four hundred fifty million and one billion.

Nutritionally speaking, malnutrition is simply a state in which the body does not receive what it needs from the food ingested. One of the causes of malnutrition is too little food but it can also be caused by an imbalance of one or more kinds of nutrients, such as protein. Severe malnutrition— what physicians refer to as "clinical malnutrition"—encompasses a number of familiar and not-so-familiar diseases including pellagra, scurvy, and rickets. Since children require about twice as much protein and calories as adults, malnutrition strikes them more often. The most severe form of malnutrition, protein-calorie malnutrition, is a childhood syndrome. It is manifested in two diseases, *kwashiorkor,* which is caused by a protein deficiency, and *marasmus,* which is caused by a calorie deficiency. If untreated, both can and do lead to death. According to one estimate, ten to twenty million young children in developing countries have severe forms of *kwashiorkor* or *marasmus* at any one time.

For most of the world's hungry, though, malnutrition takes the less dramatic form referred to by physicians as "subclinical malnutrition." Victims show symptoms such as apathy, weakness, low weight, short stature, inability to handle stress, overexcitability, and lowered resistance to disease. Again, children are affected more severely and frequently than adults. Subclinical malnutrition isn't directly lethal, but it sharply raises the odds of early death from other causes. In one survey carried out in Central and South America, 57 percent of all child deaths were found to be associated with nutritional deficiencies. The symptoms of *kwashiorkor* and *marasmus* are unmistakable, but parents of malnourished children often fail to realize that their youngsters' apathy or weakness is due to diet. "Neither the children nor their parents realize they are sick because they do not know what it is to be well," says Dr. Sohan Manocha, a biologist at Emory University in Atlanta, and author of *Nutrition and Our Overpopulated Planet.* Based on surveys in various underdeveloped countries, Worldwatch Institute estimates that subclinical malnutrition affects as many as one-half to two-thirds of the children in underdeveloped countries.

Kwashiorkor and *marasmus* are almost unknown in developed countries, but milder forms of malnutrition are found even in this land of drive-in hamburger stands and pizza parlors. Most of our malnourished are members of minority ethnic groups, migrant workers, or the aged. In a large-scale (eighty-three thousand people) survey undertaken in the years 1968–1970 by the U.S. Department of Health, Education, and Welfare in low-income areas in selected states, protein intake fell under 50 percent of accepted dietary standards for 9 percent of all households, 9 percent of infants, 11 percent of adolescents, and 21 percent of the aged. Protein intake was between 50 percent and 69 percent of recommended standards in an additional 19 percent of all households, 3 percent of infants, 12 percent of adolescents, and 18 percent of the aged. Calorie intake was below the 50 percent standard by an even wider margin, ranging from 14 percent for infants up to 25 percent for the aged. The figures are for five states: Texas, Louisiana, New York, Kentucky, and Michigan.

In recent years, the presence of large numbers of malnourished children in developing countries and smaller numbers in developed countries has taken on a new and frightening significance. Intelligence in children, new studies indicate, is linked with malnutrition, particularly protein malnutrition, not only in the children themselves but in their mothers as well.

The link between malnutrition and intelligence was first suggested in the 1960's by a team of researchers headed by Dr. Stephen Zamenhof at the University of California at Los Angeles. Zamenhof fed pregnant rats a diet deficient in protein, producing a reduced number of brain cells in the offspring. The protein content of the brain cells was about 20 percent lower than normal. About the same time, another research team, this one headed by Drs. J. K. Stephan and Bacon F. Chow of the Johns Hopkins University School of Medicine and Dr. Myron Winick of Cornell University Medical College, showed that malnourished rat mothers had smaller, lighter placentas—the temporary mass of tissue that provides the fetus with nourishment—than mothers that were adequately fed. When the offspring from these mothers were born, they, too, had fewer brain cells.

The significance of this work lies in the makeup and development of the human brain. It is composed of cells called neurons—some eleven billion in a normal adult—plus many other millions of cells called neuroglia that support and connect the neurons. The neuroglia continue to increase after birth, but not the neurons. Humans are born with their full complement of neurons; those that die are never replaced. A baby born with fewer neurons than normal never acquires more. Other research has shown that there are two spurts of growth in the human brain. One occurs during the fifteenth to twentieth weeks of pregnancy and involves neuron multiplication primarily. The other begins about the twenty-fifth week of pregnancy and continues until the second year after the child is born. Most of the growth in the second spurt, however, ceases after the first few months of life. The second spurt primarily involves the neuroglia. If anything—maternal malnutrition, for instance—interrupts this carefully timed sequence, the damage is not repaired. Neurons that did not grow between the fifteenth and twentieth weeks of pregnancy never grow. Neuroglia that did not grow between the twenty-fifth week and the beginning of the second year after birth never grow.

"The brain never gets another chance," says Myron Winick.

Most brain studies have been carried out with rats for the obvious reason that it is possible to do work with rats, such as killing them and weighing their brains, that it is impossible to do with human beings. A few studies have been performed on human brains, however, and they bear out the rat studies. Working with the brains of nineteen Chilean children who had died accidentally, Myron Winick and a colleague at Cornell University Medical College, Dr. Pedro Rosso, found that the brains of the ten well-nourished children had the same number of cells as those of well-nourished U.S. children. But the brains of the nine severely malnourished children had a drastically lower number of brain cells. Three of the children weighed less than 2,000 grams (4.4 pounds) at birth; they had a 60 percent reduction in brain cells. Drs. Donald F. Caldwell and John A. Churchill of Wayne State University in Detroit charted the dietary history of poor black women during their pregnancies. A group that ate fewer than 50 grams of protein per day (the recommended daily allowance for pregnant women is 76 grams) bore babies that weighed less and had smaller skulls than the group eating more than 70 grams.

But does a reduction in brain cells affect intelligence?

Research suggests that it does. In a study that appeared in 1967, Caldwell and Churchill, the researchers mentioned above, showed that the learning ability of rats was significantly lowered when the mother received a low-protein diet during the last part of her pregnancy. Work with human infants supports the rat data. A Canadian study of 502 children shows that those with a very low birth weight—a common sign of malnutrition—consistently scored lower than children of normal birth weight on intelligence tests. Another study, this one carried out in Scotland at the University of Edinburgh, shows that what are called "small for date" children, as opposed to premature children, had a much higher incidence of mental and physical handicaps than normal children.

The intelligence gap between malnourished children and children fed a normal diet has been clarified in another group of studies. Two physicians who undertook a five-year study of forty-two children of mixed black, white, and Malaysian blood in South Africa reported a mean intelligence of 70.86 for the underfed group, a mean of 93.48 for the well-fed group. In Yugoslavia, two physicians found a difference of 18 points between the average I.Q. scores of malnourished Serbian children and those of well-fed children from the same background.

The most depressing fact that emerges from this work is that an adequate diet after birth may not make up for the brain damage done before birth. Protein deprivation, in particular, has long-term effects. Stephen Zamenhof, the University of California researcher whose work first uncovered the link between malnutrition and intelligence, put female rats on a protein-deficient diet for one month prior to mating. After mating with well-fed males, the females were kept on the same deficient diet until their offspring were born. The female offspring were well fed either from birth or from the time of weaning and through their subsequent adulthood, mating, and pregnancies. Their offspring—the second generation of rats—were also well fed. Nevertheless, these well-fed young rats, the grandchildren of the malnourished rats, had smaller brains and fewer brain cells than normal rats.

"The damaging effects of prenatal malnutrition can operate *for at least two generations*," notes Dr. Elie Sheneour in his book *The Malnourished Mind*.

Happily, some recent studies hold out hope for malnourished minds. Dr. Joaquin Cravioto of Mexico, a noted researcher in the field of nutrition, provided children hospitalized for malnutrition with extra sources of environmental stimulation—toys, talks with friendly adults, and so forth. Preliminary findings from his work indicate that these children show a much more rapid cognitive development than normally occurs with young patients being treated for malnutrition. Dr. Slavka Frankova of Poland achieved similar results with malnourished young laboratory animals by raising them with an adult animal that provided an extra source of stimulation. Perhaps the most significant study was carried out by a group headed by Myron Winick. They found that malnourished Korean children adopted by American families at an average age of eighteen months scored as well on intelligence tests as the average U.S. child. But the same study showed that Korean children who were well fed in infancy scored *better* than the average U.S. child, indicating that the malnourished children lag somewhat behind the well fed even after the former make up much of the intellectual gap.

As some of these studies indicate, lack of protein in the developing fetus and infant is implicated directly in lessened intellectual ability. Too little protein produces adverse effects in rats and children even if calorie intake is adequate.

Protein is of such vital importance in early life because it is the substance required by the body to build cells, as well as to maintain and repair them. Contrary to popular opinion, babies and pregnant women need much more protein than athletes. A severe illness or operation also calls for extra protein. The basic unit of protein is the amino acid, twenty-two of which are known. Of these, eleven cannot be made

ESSENTIAL AMINO ACID PATTERNS IN
FAO REFERENCE PROTEIN AND HEN'S EGG

AMINO ACID	FAO REFERENCE PROTEIN	HEN'S EGG
Isoleucine	4.2	6.8
Leucine	4.8	9.0
Lysine	4.2	6.3
Phenylalanine	2.8	6.0
Tyrosine	2.8	4.4
Cystine *	2.0	2.3
Methionine *	2.2	3.1
Threonine	2.8	5.0
Tryptophan	1.4	1.7
Valine	4.2	7.4

* Cystine and methionine, the sulphur-containing amino acids, are often considered together because cystine can be formed from methionine. The combined level for cystine and methionine in the FAO reference protein is 4.2, in the egg, 5.4.

by the body; they can be obtained only from food. These proteins are known as the essential amino acids. They are: isoleucine, leucine, lysine, methionine, cystine, phenylalanine, threonine, tryptophan, valine, histidine, and tyrosine. Beyond the fetal stage, the body can make cystine from methionine and tyrosine from phenylalanine, so cystine and tyrosine are not needed if the body gets enough methionine and phenylalanine. The histidine requirement is so low for healthy adults it has little practical importance. This leaves eight major essential amino acids.

The essential amino acids must be given to the body in

certain amounts. A protein source that supplies all the essential amino acids in the correct amounts is called a "complete protein." If you eat a modest amount of it, it supplies all the protein you need. "Incomplete proteins" are low in one or more of the amino acids. To make it easy to check the amino acid levels in foods, the Food and Agricultural Organization (FAO) of the United Nations has devised a hypothetical reference or "perfect protein." To find out whether a protein source is complete, you measure the amount of its essential amino acids against those in the perfect protein. If the amount of any amino acid falls significantly short of that in the reference protein, it becomes what is called the 'limiting' amino acid. It literally "limits" the utilization of the protein taken into the body. In practice, this means that if you are eating a food in which one particular amino acid is short—let's say it's one-half that of the perfect protein's—you get only half the benefit of *all* the essential amino acids.

Foods containing the balance of essential amino acids closest to what the body needs have a high "biological value" or BV, a term frequently used by nutritionists. BV is a measurement of the amount of nitrogen, an important component of protein, retained and absorbed by the body from a particular food. It is expressed as a percentage. The food with the highest biological value is the egg, which has a BV of 96. Milk is 83, meat (including fish and poultry), 80. Theoretically, you could live on eggs, milk, or meat alone. Trailing behind eggs, milk, and meat are the plant protein sources: wheat germ, soybeans, rice, wheat, potatoes, oats, corn, peanuts, beans, and peas, in that order. Biological value isn't related to the amount of protein. Soybeans, for instance, contain much more protein, gram for gram, than potatoes; potatoes, however, more closely approximate the egg in amino acid composition, thus raising their BV. The lowly spud, in fact, is one of our best sources of vegetable protein, with a highly respectable BV of 78. The protein content is only about 4 percent, but efforts are currently underway at the University of Minnesota to increase it. Dr. Sharon Desborough, who is in charge of the research, has already produced an experimental potato with a 20 percent protein content. Most plant protein sources are low in one or more of the amino acids lysine, tryptophan, methionine, and threonine. To bring a plant protein diet up to the BV of a meat diet, one has to choose a judicious mixture of plant protein sources with complementary amino acids.

One caution: The mixture of plant protein sources must be taken together; waiting even a few hours to add one of the supplementary plants results in the waste of much of the protein.

In the United States, with its ample meat and dairy resources, we prefer to get most of our proteins from meat, milk, cheese, and eggs. They are our traditional sources of protein. As pointed out earlier, many nutritionists believe most Americans eat too much protein. According to the FAO/ WHO, an average-size adult male (about 154 pounds) needs 33 grams of protein per day, an average-size adult female (about 143 pounds), 30 grams a day. The same requirement is set by the National Research Council of the National Academy of Sciences, a private U.S. scientific organization. Pregnant women should add about 10 grams of protein per day. Children's needs exceed adults and decrease with age. To find the FAO/WHO-NRC/NAS requirement for each age, you multiply weight by a sliding scale that begins with 2.2 grams of protein per kilogram of body weight for infants and drops to 0.5 grams for adults.

If you determine the recommended daily protein requirement for your age and weight, however, that doesn't translate into the amount of protein you should actually eat. Individual protein requirements vary and so does the efficiency of utilization for various food sources, mandating a much higher consumption of protein to make sure the recommended *requirement* is met for every individual in an age and weight group. This higher level is known as the recommended *allowance*. The allowance is what we are actually supposed to eat. For an adult male of average size, for instance, the NAS/ NRC-FAO/WHO daily protein allowance is 56 grams. An adult can translate his or her weight into the allowance by multiplying weight in kilograms by 0.8. The same standard applies to both meat-based diets and diets based on mixtures of vegetable proteins.

Some nutritionists quarrel with these standards, some claiming they are too high, others too low. *Human Nutrition*, a classic text in the field, takes the middle ground. The FAO/WHO standard, it says, "appears to provide a margin of safety when applied to healthy persons with adequate caloric intake, normally vigorous, and living in a temperate climate." The average American diet, with its 99 grams of protein, however, exceeds all recommended standards. The excess proteins we eat do not make us healthy; there is some evidence that an excess of animal protein is distinctly unhealthy (see Chapter 14). Also, excess protein can make you fat. After the protein we eat has been converted into amino acids to meet the needs of the body, the excess amino acids are used for energy and heat or, if not needed for that purpose, converted into fat and stored. The part of the protein molecule that cannot be used for energy is excreted.

People in Western Europe and a few other areas of the world eat a diet much like ours, but the greater part of the world gets most of its protein and calories from plants. Rice and wheat are the staple foods of more than two-thirds of the human population, rice in the tropics, wheat in temperate and dry regions. Corn, sorghum, oats, rye, and millet are popular in various areas. Cereal grains like those named are not a balanced source of protein, being low in lysine (with the exception of hybrids such as Opaque 2 corn), but they provide a diet that is adequate in biologic value if supplemented with a high-lysine plant, such as beans, or small amounts of meat and dairy foods. Even wheat alone can sustain life, although it will be life at a lower level of efficiency and with the possible hazard of damage to the intellect if an all-wheat diet is eaten by a pregnant woman or an infant. But millions of people live long, healthy lives eating mixtures of grains and beans, or grains with small amounts of meat or cheese. Since meat and dairy foods are expensive to produce, nutritionists and economists alike now believe that plants will continue to be the source of both proteins and calories for most of the people on this planet.

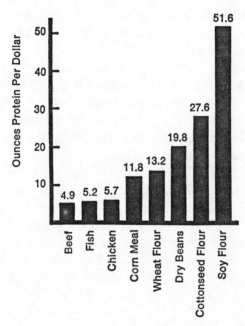

Relative Costs of Utilizable Protein
Source: Bird, 1972.

A graphic illustration of the dependence of the world's developing countries on plant proteins was furnished by a report that appeared in *The New York Times* on December 12, 1976. It related that people in Haiti, which is one of the poorest nations in this hemisphere, were selling the cans of beef chunks donated to them by an American church organization for rice, beans, and corn, the staples of their diet. The cans of beef subsequently appeared on shelves of stores in Port-au-Prince, the capital, where they were purchased by wealthier Haitians. From the standpoint of the poor Haitians, selling the beef for a larger portion of grains was "the intelligent thing for them to do," according to a spokesman for the American church. One ration of beef, he pointed out, could be exchanged for five rations of rice.

But will there be enough plant protein to go around in the future to provide the optimum mixture of amino acids? Will there be enough of even one fairly good source of plant protein like wheat or rice?

The world population explosion indicates that the answer to both questions may be no. Consider a few figures plucked from various reports. The world now has about 3.92 billion people. The population is growing at the rate of about 64 million people per year. By the year 2000, the world population is expected to be about 6 billion. Asia and Latin America, between them, will have about 80 percent of the world's population by then; India alone will have about 1 billion people. Some of the more pessimistic prophecies project 50 billion people on earth less than a century from now. The United Nations, however, forecasts a stabilization at between 10 and 12.5 billion by 2025 to 2035. If this more conservative estimate is borne out, there will be two and one-half to three times the population that now exists on earth fifty years from now!

Food supplies, many agricultural experts fear, cannot keep up with the population. "We are attempting to deal with finite limitations of resources in the face of what appears to be infinite human potential for procreation," said J. George Harrar, president emeritus of the Rockefeller Foundation, in 1974.

Coming from Harrar, the comment is especially significant because the Rockefeller Foundation is the organization that financed much of the international research that led to the "Green Revolution." In the 1960's, the Green Revolution promised to rid the world of hunger. By planting hardy new high-yield varieties of wheat and rice, developing countries expanded their yields of those two vital crops by about 35 percent per acre. The new varieties were developed at international agricultural research centers staffed by scientists from many countries, including the United States. American scientist Norman Borlaug won a Nobel Peace Prize for his efforts in developing high-yielding wheat at CIMMYT (the initial letters stand for "International Maize and Wheat Improvement Center" in Spanish), a research center in Mexico. But the Green Revolution may have run its course, at least as it was initially conceived. The new varieties of wheat and rice require large amounts of fertilizer, pesticides, and water. Fertilizer and pesticides are made from fossil fuels—oil and gas. With the rise in the price of oil, the price of fertilizer and pesticides has leaped upward. Economists predict developing countries will be unable to pay for the fertilizer they will need in the future—even if it is available. The two million excess deaths attributed to hunger in the early 1970's indicate that the new high-yielding varieties are already failing to keep pace with the growing population.

The United States is directly involved in the world food picture because our country and Canada are the leading exporters of the grain on which much of the world depends for both calories and protein. Australia and New Zealand also export large amounts of grain. Our grain customers include Russia and Western European countries, Japan, China, India, Africa, and South America. Or, in other words, most of the world. We were not always the world's number one grain supplier. Before World War I, according to figures compiled by Lester Brown of Worldwatch, many developing countries exported grain to developed countries. In recent decades, however, these poor nations have been unable to grow enough food to feed their own rising populations, much less export it. Some food experts predict a grain deficit in the developing countries of some 100 million tons a year by the mid-1980's. Developed countries like Russia will probably always be able to pay for extra grain they need in times of temporary shortages, but how will the developing countries pay the bill for the grain they'll need to feed a vastly increased population? Will richer nations donate the food? And if they do, will a situation be set up in which more and more people demand more and more grain, until even the grain exporters can't make up the deficit?

In the mid-1970's, predictions concerning food and population were invariably gloomy. As the 1980's approach, however, two trends seem to have revived hopes. The first is a lowering of the birth rate in some developing countries, the second another upswing in agricultural yields in developing countries. A "cautiously optimistic" report issued by the National Research Council in mid-1977 concludes that the worst aspects of widespread hunger and malnutrition may be overcome within one generation—twenty to twenty-five years—given certain conditions. The conditions: a doubling of food production by developing countries by the end of the century, better use of existing food supplies, lowering of birth rates, what the NRC calls "political will," and the enlistment of the efforts of developed countries, particularly the United States, in the search for better crops and methods of storing them. The last condition is already on the way to implementation. The new emphasis at the international research centers where the Green Revolution began is on crops that need little or no pesticides or fertilizer. Even in the United States, research attention is turning to new ways to make plants yield well without fertilizer or pesticides.

One road to high yields without fertilizer that is being explored by a number of research groups today is biological nitrogen fixation. Some plants, principally beans among the agricultural crops, need less nitrogen fertilizer, the major fertilizer used throughout the world, because they are able to change nitrogen from the atmosphere into compounds usable by plants. They do it by means of bacteria and algae that grow naturally on their roots. Throughout the history of agriculture, farmers have taken advantage of this natural system by plowing under beans at the end of a growing season and then sowing grains. Grains, which cannot convert atmospheric nitrogen, take up nitrogen transferred to the soil by the bean plants. U.S. researchers are trying to improve nitrogen fixation in such plants as soybeans and transfer the capacity to grains.

The present trends in food production and population are encouraging, but the balance between the two could be upset by a number of factors. One is the weather. In case you hadn't noticed it, the weather has been up to some strange tricks lately. Beginning in 1972, a series of droughts and floods has affected many parts of the world, including the United States. Some of our north central states had their driest growing season in a quarter-century in 1973 and 1974. Rainfall is up again but the 1976–77 winter was the coldest in memory for many people in this country. Is this bad weather just an aberration in the normal patterns? No, say many climatologists, who predict the weather is going to get worse, with inevitable consequences for food production. Dr. Iben Browning, a climatologist who serves as a consultant to many U.S. business firms, believes the last fifty years were abnormally warm and that we are on our way into a long period of cold weather like that which prevailed during most of our history. " 'Normal' is climate that is just terrible," he says. If Browning is right (and there are many climatologists who share his viewpoint), the grain, corn, and soybean belts will shift southward, taking some large growing areas out of production.

The United States would suffer less than most of the major agricultural countries in the event of such a shift because our grain and soybeans can be grown in the southern part of the country if the north becomes too cold. But there would be serious consequences in countries like Russia, where the grain belt is already in the south.

With respect to the world food shortage, the United States is in a fortunate position in many ways. All predictions indicate our farmers have the climate, land, agricultural know-how, and political support to raise enough food for our own population as well as for many of the world's needy for the foreseeable future. Our agricultural resources are better than the Arabs' oil resources—and, unlike oil, they won't run out. This doesn't mean, however, that a much larger world population, with its attendant food problems, will not affect us. Our present diet, with its emphasis on overconsumption of expensive, protein-rich foods like meat and dairy products, seems destined for a change. What form will it take? We will not switch to a vegetable-based diet similar to that of the developing countries, but we will eat less meat and dairy products and more nontraditional forms of protein like soybeans and vegetables that are unknown here today. Even our traditional forms of protein will undergo a change. Grain-fed beef may become a rarity in the year 2000, with most beef being reared to market size on forage. Grain will have more protein and lysine. And we will make much better use of nutritious food wastes which today end up in our sewage systems.

The major reason for the switch to such a diet is economic. Traditional sources of protein cost more today than they used to; from all indications, they will continue to rise in price. Most nontraditional sources of protein, on the other hand, are fairly cheap. A 1976 study by Dr. David Pimentel and Marcia Pimentel of Cornell University shows that plant protein costs one-third to one-half as much as animal protein. One way to keep food costs a constant proportion of our income in a period of rising prices (less than 20 percent of our income today), points out Dr. William W. Gallimore, a U.S. Department of Agriculture economist, is to put more of the cheaper alternative sources of protein into our diet. "Consumers in the United States now face the possibility of changing their diets or increasing the percentage of income spent on foods," says Gallimore. Americans, he believes, will accept a change in diet. A soy-ground beef blend offered in many supermarkets in the mid-1970's sold well as long as the price of the blend was sufficiently lower than that of the all-meat product. The shift to a diet with less meat and more alternative sources of protein has another economic advantage, too. It would allow us to divert grain away from our livestock and into exports, thus shoring up our balance of payments, which has been affected adversely by oil imports.

The latter argument is a particularly attractive one, because it allows us not only to help our deteriorating balance of payments but also to supply more food grains to people, not animals, a move which some scientists now urge for humanitarian reasons. Our present per capita consumption of grain is about two thousand pounds, compared to only around four hundred pounds in the ·developing countries. You didn't eat two thousand pounds of grain last year? Of course you didn't; most of our grain goes to animals. In the developing countries, most grain goes to people. It takes three to ten pounds of grain, an average of seven pounds, to produce one pound of meat under the U.S. system. Grain isn't the only human food fed to animals. Over 90 percent of the soybeans consumed in this country are eaten not by people but by animals, principally pigs and chickens. Soybeans are 34 to 37 percent protein, the highest protein content of any plant with the possible exception of a few obscure species.

Animals are also profligate consumers of other resources. It takes ten times as much water, three to six times as much space, and ten to twenty times as much energy to raise livestock as it does to raise plant protein. The Pimentel study shows that animal protein sources require far more energy per gram of protein produced than plant protein sources. Forage-fed beef, the most efficient animal protein source, has a ratio of fossil energy input to protein output of ten to one; soybeans, the most efficient plant protein source, have a two to one ratio. Feedlot beef, the least efficient animal protein source, has a ratio of fossil energy input to protein output of seventy-seven to one! By comparison, the least efficient plant protein source, rice, has a ratio of ten to one, the same as that of the most efficient animal protein source.

How long can we afford to acquire our protein in a way that not only consumes an inordinate proportion of the world's resources but also threatens to send our food budgets over the 20 percent of income mark in the future?

A number of experts agree that a change is imminent. Feedlot operator Kenneth Monfort, quoted in *The Wall Street Journal*, says beef consumption "will drop dramatically" in the future. An even more startling prediction comes from

WORLD CEREAL CONSUMPTION,
AS FOOD AND FEED

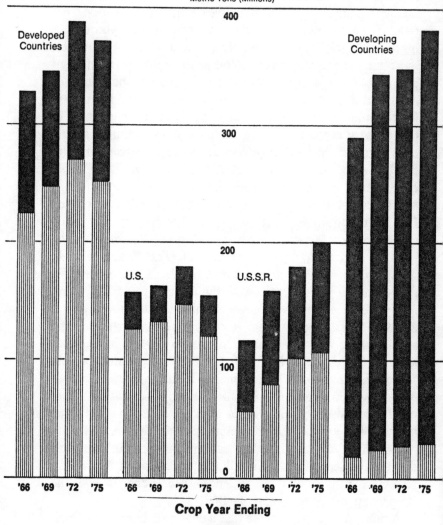

Source: Rockefeller Foundation, 1975

Dr. D. M. Sahasrabudhe of Canada's Food Research Institute: "We shall be short of meat products in the future. Meat will become a delicacy and very expensive." Alternative protein sources, other experts believe, will replace some of the meat in our diet. "In another several decades I believe we will be obtaining one-half to two-thirds of our food-grade protein from plant-derived sources," says Department of Agriculture nutritionist Kermit Bird, also in *The Wall Street Journal*. The change may come even earlier. Drs. R. W. Forsythe and Ernest J. Briskey, two food scientists with the Campbell Soup Company, predicted in 1976 that consumption of soybean-based protein foods would increase 20 to 25 percent in the next five years. Alternative forms of protein that will be consumed in larger quantities by the year 2000, they claim, are single-celled protein sources like yeast, fish protein concentrate, and waste meat products that today are fed to pets or thrown away.

Some alternative forms of protein are already available in the supermarket or health food store and all of them are under investigation at research institutions here and abroad. Some will undoubtedly find a secure place on our dinner table. The good news about these foods is that they are good. With the exception of a few products like meat "analogs," the new imitation meat now being sold in many supermarkets, alternative forms of protein are traditional forms of protein in some areas of the world. People like them. "Dry beans a new form of protein?" a friend born in Spain cried, shocked, when I described one of my topics. "There's nothing new about beans in Spain."

Someday there won't be anything new about most of the foods described in this book.

2

THE CINDERELLA BEAN

"By George, they've done it! This fillet
steak tastes exactly like soya beans!"

—CARTOON CAPTION
New Scientist

THE TWO HUNDRED-ACRE EXPERIMENTAL FARM OPERATED
by the University of Illinois looks much like any flourishing
Illinois farm except for one missing element: the farmhouse.
The barn and other outbuildings are grouped together in one
area, surrounded by a perfectly flat expanse covered with neat
rows of crops. A cow moos from the direction of the barn.
I was there on a hot August day to see one crop in particular:
soybeans. Like most urban Americans, I had never actually
seen the plant that is our number one cash crop, our leading
agricultural export, the source of most of our edible oil and
animal feed protein, the world's richest source of plant protein
and—the real reason for my interest in the soybean—probably
the major source of unconventional protein for humans in
years to come. Dr. Richard L. Cooper, director of the U.S.
Regional Soybean Laboratory, a U.S. Department of Agriculture research operation housed at the university's Urbana
campus, agreed to repair the lapse in my soybean knowledge
with a tour of the laboratory's soybean fields.

On the way to the farm, we drove through the main part
of the campus, right in the middle of which is a sizable vege-

table garden. "The oldest experimental plots of corn and soy-beans in the country," explained Cooper, a tall, suntanned man who wore jeans and boots. "They were planted in 1878. The building next to it was built underground so it won't shade the crops."

The experimental farm itself is just across the highway from the campus. Cooper stopped the car and walked over to a plot of leafy, three-foot-high soybean plants. He reached down and took off a fuzzy green pod about two inches long. Inside it were two yellowish seeds, each about the size and shape of a pea. At this point, he explained, the plant is unripe; only in the fall, when the pods are brown and dry, will the soybeans be harvested. He held back some of the heavy foliage. Every plant was full of pods. The world's highest soybean yields are achieved in this flat area of central Illinois.

"Ninety-two bushels an acre is my record," said Cooper. He grew them, he explained, on a private farm a few miles away. Yields at the experimental farm are in the eighties. A few farmers in the area have achieved one hundred bushels an acre on occasion.

A young French grower who accompanied us on the tour looked impressed. Soybeans are grown throughout the world in temperate climates with wet summers and cold winters, but anything over 30 bushels an acre is a dream yield for most of the world's farmers. The average world yield is 18.2 bushels. For the United States, the average yield is 28.4 bushels; for Illinois, 35 bushels, the highest state average in the nation. Not surprisingly, Illinois is the nation's leading producer of soybeans. The runners-up, in order, are Iowa, Indiana, Missouri, Minnesota, Arkansas, and Ohio.

"What makes Illinois such a good place to grow soybeans?" I asked as we tramped through closely planted ranks of soy-beans to another field. "Soil," answered Cooper. "And water. Management is important, too, but we have good soil and good rainfall and water-holding capacity and that attracts the best managers."

Cooper's special interest is finding an even higher-yielding

AVERAGE AND RECORD YIELDS
FOR SOME FOOD CROPS

FOOD	AVERAGE, 1974	RECORD
Corn (bushel per acre)	72	307
Wheat (bushel per acre)	28	216
Soybeans (bushel per acre)	24	110
Sorghum (bushel per acre)	45	320
Oats (bushel per acre)	48	296
Barley (bushel per acre)	38	212
Potatoes (bushel per acre)	420	1400
Sugar beets (ton per acre)	19	54

Source: S. H. Wittwer, *Science*, Vol. 188, May 9, 1975, Copyright 1975, AAAS.

soybean that will flourish not just in fertile central Illinois but throughout the northern United States and possibly in other temperate areas as well. The benefits of a higher-yielding soybean would be huge, not only for the United States, but for other areas of the world. Soybeans contain 34 to 37 percent protein, which makes them the best source of vegetable protein in the world, with the possible exception of a few obscure plants such as the winged bean (see Chapter 8). For the past few decades, the United States has been the world's leading producer and exporter of the nutritious bean. We grow three-fourths of the world's soybeans, one-third of which we export. Our best customers are Western Europe and Japan. In Europe, as in this country, most soybeans are used as a protein source in animal feeds, but in Asia soybeans are a major source of protein for people. Dr. R. W. Howell, head of the Department of Agronomy at the University of Illinois, estimates that a 10 percent rise in the average U.S. yield of 28.4 bushels of soybeans per acre would produce enough additional protein to feed some seventy million people.

Is a rise in soybean yields likely? Cooper figures it is just over the flat Illinois horizon. On our tour of the experimental farm, we came to a thickly planted plot where the plants are much shorter than the soybeans around them. In spite of the crowding, the diminutive plants were leaning over as if a high wind were blowing, but on this hot afternoon there was no wind. Cooper stopped in front of the little plants, which looked, to my eyes, like weaklings compared to all the three-foot-tall plants around them. "I've used this field for semi-dwarf plants," he said. "As you'll see, there are not so many pods on any one plant, but you can plant a lot more of them. Look at the number of pods." He picked up a plant that had fallen over and to my surprise it was full of pods. "We call it lodging when the plants lie down like that," he said in answer to my question. "Corsoy—the semidwarf variety we're looking at here—is lodging resistant. That means it yields well in spite of lodging. Placing the plants closely together also increases yield. But what I really want to do is develop a plant that yields well and doesn't lodge. I think lodging is a barrier to yield."

He led the way to another plot of diminutive plants, this time with comparatively little lodging. "We have three acres of this under cultivation," he said. "It got seventy bushels the first year and eighty bushels the second. If it does well this year, I may release it to growers. What will we call it? Well, my daughter calls it Elf, but I don't know if that's the name we'll use." Walking back across the fields, he talked about some of the difficulties in producing a new variety of soybeans. "Since 1969, we've made at least one hundred fifty crosses a year to produce the high-yielding semidwarf. It probably comes to something like sixteen hundred crosses." He added, "The trick of successful breeding is knowing what to throw out."

Cooper has ample material to draw on for crossing soybeans. A few doors down the hall from his office in the U.S. Regional Soybean Laboratory, a warren of offices and laboratories in the basement of an old building on the University of Illinois campus, is a temperature-controlled room called the Soybean Bank. It contains some five thousand different varieties of soybean seeds from all over the world. The bank receives several hundred requests for seeds each year from growers. Most of the seeds at the Illinois bank are adapted to a northern climate; seeds adapted to warmer climates are maintained at a facility in Stoneville, Mississippi. In all, according to Cooper, there are some ten thousand to fifteen thousand varieties of soybeans in the world, about fifty of which are grown in the United States. New varieties are released from time to time to meet various needs. One possibility is a super-high protein variety. The soybean bank has several seeds in which the protein content is around 50 percent. Yields are low, however, and extensive crossing will be needed to combine high protein traits with high yields.

"Farmers won't trade off yield for anything," Cooper noted, a refrain given to me by a number of agricultural researchers.

There are other routes to the high-yielding soybean besides breeding. The soybean, known as *Glycine max* (L.) Merrill to scientists, is a member of the legume family, all of which, as mentioned in Chapter 1, have the ability to make their own nitrogen fertilizer. The soybean does not actually make its own fertilizer. The plant furnishes hydrogen to bacteria of the genus *Rhizobium*, which live on its roots. The bacteria, in turn, use the hydrogen to extract nitrogen from the air and fix it as ammonia in the soil. The ammonia nourishes the soybeans. But while the soybean can make its own nitrogen via this process, it does not do a particularly good job of it. Some plants, including marsh grasses, do it better. A number of researchers are now unraveling the intricacies of biological nitrogen fixation in hopes of improving it in plants that already use it and transferring it to ones that do not.

One of the most promising recent developments in this area has been achieved at Oregon State University, where researchers have switched the type of bacteria that usually lives on soybean roots to one that does a more efficient job. The strain of *Rhizobium* associated with soybeans in nature wastes almost half the hydrogen supplied to it by the plant. Since bacteria use up about 12 percent of the soybean's total hydrogen output, this means about 6 percent of the soybean's total hydrogen output is lost. The new strains of *Rhizobium* being used at Oregon State apparently possess an enzyme—a protein that promotes life processes—that enables them to retain almost all the hydrogen supplied to them. Dr. Harold J. Evans of the Oregon State team estimates the discovery might increase soybean yields by as much as 10 percent.

Before I left Urbana's soybean fields, Richard Cooper gave me a copy of the proceedings of a symposium in which he had participated entitled, "Soybeans: The Wonder Bean." The wondrous nature of the soybean is a common theme among agricultural researchers, usually a plain-spoken group. In the pile of scientific papers on soybeans I have in my study, the bean is lyrically referred to as the "miracle bean," the "dream bean," the "golden bean," and (my favorite) the "Cinderella bean." Reading through these papers, it is easy to see why hardheaded researchers, farmers, and food processors wax so poetic over the little yellow bean. It really is a Cinderella bean, at least in this country. The biggest success story in American agriculture, the soybean has risen, metaphorically, from rags to riches in less than half a century. The tale goes something like this:

In the late eighteenth century, some small yellowish beans native to the northern region of China known as Manchuria were brought to this country. The bean was—and is—the basis of a number of traditional foods, some of them staples, in the country of its origin. When the beans were planted in temperate regions similar to those where it flourished in Manchuria, they did well. The first mention of the soybean in this country appears in an agricultural publication in 1804. But what to do with the beans? A few farmers' wives undoubtedly tried to prepare them and found the beans remained hard even after hours of cooking. So farmers simply used the whole plant—leaves, stem, pods, and beans—as hay, which they fed to livestock. For a while, in fact, soybeans were called "haybeans" here. Dr. M. Wayne Adams, a Michigan State University agricultural researcher, remembers putting up "haybean" hay as a boy.

In the first few decades of the twentieth century, far more lucrative uses were found for the soybean, beginning with its oil. Soybeans are technically known as "oilseeds" because of the high oil content of the bean. In addition to the 34 to 37 percent protein, soybeans contain about 20 percent oil. Both substances are neatly packaged in separate "bodies" that can be seen under a powerful microscope. In 1921, a firm in Decatur, Ill., already a prime growing area for soybeans, offered farmers the unheard-of sum of $1.35 for a bushel of soybeans (at this writing, soybeans are selling for about $9.00 a bushel). Like other processors at that time, the Decatur firm was interested exclusively in the oil. Removing soybean oil, however, leaves behind a crushed, defatted bean. At first this by-product was sold for fertilizer, but before long oil processors began grinding up the beans into a crude meal, which farmers used as a cheap animal feed. Cows, pigs, and chickens thrived on it. New analytical techniques that became available in the 1950's revealed that the plant that was once used for hay has a bean containing the richest source of protein among the vegetables.

The fact that the soybean can be used for both oil and protein is the secret of its success here. A protein content of 34 to 37 percent is very high for a seed, but what makes the soybean even more appealing is the composition of that protein. For a protein source to be completely available to the body, it must contain certain amounts of each of the eight essential amino acids. Soybeans are significantly short in only one essential amino acid, methionine, and are generously supplied with another, lysine, in which cereal grains such as wheat and corn are low. Several nutritional measures used by scientists underscore the high value of soybeans as a protein source. The Protein Efficiency Ratio, PER for short, is a measure based on the grams of weight gained by young laboratory rats per gram of protein ingested. PER's for meat and eggs range from 2.5 to 3.5, for cereal grains from 1.0 to 2.0, for most beans

PER's OF VARIOUS PROTEIN SOURCES

Egg—whole	3.8
Pork tenderloin	3.3
Beef muscle	3.2
Wheat germ	2.9
Milk—whole	2.8
Beef liver	2.7
Ham	2.7
Soybeans (heated)	2.3
Oats (rolled)	2.2
Wheat bran	2.0
Potato	2.0
Peanut	1.7

Source: Campbell *et al.*, 1959.

from 0.50 to 1.50. Soybeans have PER's ranging from 1.5 to 2.5, better than most vegetables and almost as high as some meats.

Soybeans score even better on another nutritional measure, biological value, or BV, which, as mentioned in Chapter 1, is based on the nitrogen actually used by the body. The score is expressed as a percentage. Soybeans have a BV of 75, a little below meat at 80 and eggs at 96, but above most other vegetables tested.

A vegetable protein source with this kind of nutritional value was bound to increase in value and by the early 1950's soybean meal was selling for more than soybean oil. It still is today. The oil shortage of World War II gave another big boost to soybean production and by the time hostilities were over, the soybean was America's favorite bean—but only in the eyes of farmers, food companies that use soybeans in many processed foods, and commodity traders, who make soybeans the hottest item on the Chicago Board of Trade. As far as the average American consumer goes, the soybean is almost as little known as it was in the nineteenth century. The reason for our ignorance of the Cinderella bean is simple. To most Americans, the soybean is not a particularly desirable food item. For one thing, its taste falls short of our standards. Even its advocates say the soybean is "bland" and many Americans find it unpleasantly "beany." Also, as farmers' wives found out long ago, the soybean takes a long time to prepare. Overnight soaking is a must, and the soaked beans must be cooked for several hours—sometimes as many as nine! Cooking, it should be noted, is necessary not only for softening, but also for nutritional purposes. Animal studies show that raw soybeans are only about half as nutritious as cooked soybeans because several components in the raw bean inhibit full digestion. Even cooked beans remain firmer than most beans because soybeans have practically no starch, the component that expands to make other beans soft.

Actually, soybeans are not really that unappealing. I've made a few batches of baked soybeans—they require overnight soaking and about five hours cooking—and both my husband and I found them rather pleasant, although their firmness makes them more like nuts than beans. Health food enthusiasts swear by soybeans and you will find any number of recipes in various books sold in health food stores.

But the bean that is a flop in the American kitchen is a star in the American food processing industry. It is the source of much of the oil we use in cooking and salads, and processed soybean meal is our major source of protein in animal food. Soybean processing is a flourishing industry, with some 125 plants, trade publications, and a trade organization, the American Soybean Processors Association. Most soybean processors are located in the Midwest, just a short hop from the soybean fields. Archer Daniels Midland, the nation's biggest, is in Decatur, Illinois, right in the heart of the nation's most fertile soybean area.

Processed soy products all start out in much the same way. After the beans are cleaned, they are dehulled and put through a set of rollers, which form them into flakes. In most plants, the flakes travel on to a machine called the extractor, which uses hexane, a form of petroleum, to extract the oil from the flakes. Soybean oil, although it has no protein, is a desirable product from a health standpoint because it has a high percentage of polyunsaturated fat. Saturated fats have been linked with heart disease by some researchers. If the defatted soybean flakes are destined for animal feed, they journey on to a machine that performs two jobs: It takes out the hexane and toasts the flakes. Toasted flakes are ground for meal. The hull that was removed earlier can either be added to the meal or ground up as a separate feed. Hull contains about 11 percent protein, meal 44 to 49 percent.

Processing flakes into human food requires separate machinery to comply with federal Food and Drug Administration standards. Only about a dozen U.S. processing plants produce soybean foods for human consumption. After the hexane is removed, the flakes are toasted and ground into a product called "defatted soy flour," which is about 40 percent protein. Defatted soy flour is the cheapest and most widely used soy food. It is also the starting point for two other products, soy concentrate, which is about 75 percent protein, and soy isolate, which is about 90 percent protein (both are also produced for animals). Soy concentrate is made by removing about half the carbohydrates from defatted flour, soy isolate by removing all the carbohydrates. If the processor wants full-fat flour, a soy flour that includes the oil, he omits the oil extraction step and grinds the whole toasted bean. Each form of soy food has its special uses. Full-fat flour is likely to turn up in pet foods and fried bakery goods, defatted flour in baby food and beverages, soy concentrates in breading and sausages. The most versatile product—and the most expensive—is soy isolate, which is an ingredient in a host of foods ranging from hot dogs to whipped cream, and even in some nonfoods.

The ironic aspect of the soybean's Cinderella story is that almost all the soy protein turned out by the industry ends up in the stomachs of animals, not people. Chickens are the biggest animal consumers of soybean products, followed by pigs and cattle. Only about 3 percent of our soybean crop

SOYBEAN MEAL CONSUMED
BY ANIMAL CLASSES

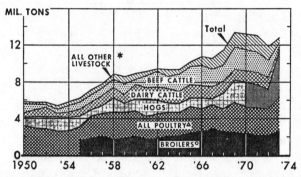

Year beginning October. *Includes sheep, horses and mules, and other livestock on farms and all livestock not on farms. △Includes chickens raised, hens and pullets, and turkeys. °Data for broilers not available prior to 1955.

Source: U.S. Department of Agriculture.

is used for human food consumed in the United States. And not only do most of our nutritious soybeans go into animal feed, but almost all of the soybeans used in human food are there not to add protein, but to improve the functional properties of processed food. In processed food, a functional ingredient is one that helps the food maintain its color, moisture content, shape, or some other nonnutritional property. Such an ingredient seldom makes up more than 5 percent of the total. Processed foods are full of functional ingredients, many derived from soybeans.

The ubiquitous lecithin, for instance, which pops up on so many labels, is a by-product of soybean oil. Lecithin improves the clarity and shine of candy, reduces the splatter of margarine, and makes baby food taste better, among other desirable functions.

If you want to get an idea of how many products soy is used in, go to your cupboard and start reading labels. Look for lecithin, hydrolized vegetable protein, most of which is soy, and anything with the word "soy"—soy flour, soy meal, and so on. When I checked my own not-too-well-stocked cupboard one day, I came up with these products with soybean ingredients: margarine, meat extender, two kinds of packaged bread, dry cereal, biscuits, dry cat food, canned dog food, dry soup, bouillon cubes, candy bars, a breakfast bar, two brands of Worcestershire sauce, barbecue sauce, lunch meat, and a can of chicken à la king. Soy is a principal ingredient in the extender and margarine, but in the others it probably plays a functional role. Incidentally, one way to check the relative amount of an ingredient in a product is to note its position on the ingredients list. The closer it is to the beginning, the more there is of it.

Just what is soy doing for all these foods? To find out, I visited the U.S. Department of Agriculture's Northern Regional Research Laboratory in Peoria, Illinois. The Peoria facility was once part of the U.S. Soybean Laboratory in Urbana but the two split their functions back in the 1930's, Urbana taking breeding research and Peoria utilization research. Dr. Walter J. Wolf is Peoria's legume utilization expert, which, in the Midwest, means primarily soybeans. He listed soy's principal functional properties recently in one of his many papers: water absorption, water holding, emulsification (preventing liquids in a mixture from separating), elasticity, fat blocking, adhesion, whipping, color control, and texture. No wonder food processors use soy! Oddly enough, though, food chemists have little idea *why* soy does all this, although the effects are generally ascribed to its protein content. The only reliable way to tell what function soy will perform in a new product, according to Wolf, is to put it in and look at the results.

In 1975, Wolf told me, U.S. processing plants produced about six hundred twenty-five million pounds of defatted soy flour, eighty million pounds of concentrates, and seventy million pounds of isolates. "Add them all up and it's three pounds per capita," he said. By way of contrast, he pointed out, we eat 128.5 pounds of beef per capita. To give me an idea of soy's functional role in foods, he took me to an area of the center where a number of consumer food products using soy had been assembled for a recent open house. A long table was covered with the items, many of them familiar to me. "There are dozens of these products here," he said, "and there are hundreds more. Mainly they are used for their functional properties."

"What on earth did processors use before soy was available?" I asked. "Egg albumen and nonfat dry milk," said Wolf. "But both are more expensive than soy."

One of the reasons why soy is cast in a functional role in processed foods, where only small amounts are used, is the taste, Wolf continued. Taste tests he and other researchers have carried out with processed soybeans indicate that many Americans think the bean has an unpleasant taste, which is variously described as "beany," "green," and "bitter." It is estimated that soy could largely replace nonfat dry milk, which is now the major functional protein ingredient in processed food, if only the offending flavor could be eliminated. Much of the soybean flavor is removed in processing, but not all of it. The product with the blandest tastes is soy isolate. If soy is used in any appreciable amount in a processed food, various flavorings and/or spices are added to mask the taste, sometimes resulting in an overflavored or overspiced item. Food chemists dream of someday removing all taste from soy, but as yet they are not even sure where the taste is. When various components of the soybean are isolated and tasted separately, they lack the unpleasant flavor of the food products.

But processed soybeans have still another problem, gas, to which various researchers refer discreetly as flatulence. "Soybeans do produce flatulence," Wolf admitted, "but only in large amounts and only with the flour." The isolate and concentrate, he claimed, cause no gas. The culprits, he and most soybean researchers believe, are two sugars, stachyose and raffinose. They cannot be digested high in the human intestines, as most food components are, so descend, undigested, to the portions of the lower intestine known as the ileum and the colon. There intestinal bacteria interact with the sugars and give off carbon dioxide, hydrogen, and other gases. Soy isolates and concentrates have little carbohydrate so they cause no gas. And, as Wolf points out, soybeans do not cause nearly as much gas as some other beans. Navy beans, for instance, cause up to two and one-half times as much gas as soybeans. The solution to the flatulence problem in processed soybeans is to remove the offending sugars without losing other nutrients, but thus far no inexpensive process for accomplishing this is available.

In spite of these drawbacks, you can find a few processed foods in your supermarket in which soy is used for its protein content, not its functional properties. In fact, you can often find packages of dried soybeans. (Look in the health food section if you can't find them with other dried beans.) None of these items was available a few years back. Probably the easiest soy-based product to obtain today in the supermarket is the "meat extenders." Introduced with great fanfare in 1973, during a period when meat was undergoing a steep price rise, the extenders first took the form of a premixed product that combined fresh hamburger with various soy extenders. It was sold right in the meat case. At first extended hamburger sold fairly well, but as meat prices declined, sales slumped. In a 1974 study, Dr. William W. Gallimore, a U.S. Department of Agriculture economist, found that when the price differential for the extended product versus ordinary hamburger dropped below twenty cents a pound, most people stopped buying it. Extenders are still around, however, although usually in a different form. In my own suburban supermarket, I found no fresh extended hamburger, but one brand of inexpensive soy-hamburger patties in the freezer case and three different brands of dry, packaged extenders, two of them made by large food companies. The frozen patties are no gourmet delight, but they are tasty enough when hot and only three-fifths the price of an equal-sized package of all-meat patties.

The frozen patty package contained no nutritional information but the dry extenders are a mine of information. One indicates that a half-and-half patty made with its product is 20 percent protein, a figure that compares favorably with the 22 percent protein content of hamburger given in *Human Nutrition*, a widely accepted textbook. A study by Muriel Happich of the U.S. Department of Agriculture indicates, however, that the PER of such patties is only 2.1, lower than that of lean hamburger. The principle behind the dry products is simple. Mix half a package of extender with half a pound of hamburger and presto! A pound of hamburger. It tastes quite good—considerably better than the frozen patties, according to my own little taste panel consisting of myself and my husband—and the savings are considerable, a package of dry extender selling for about fifty cents. But watch the labels. The most heavily advertised national brand of dry extender does not mention soy or plant protein, meaning that it extends hamburger in bulk, but cuts its protein content.

Another soy-based product readily available today is Miles Laboratories' frozen "meat analogs," as the manufacturer likes to call them. (They positively shudder at the term "imitation meat.") Sold under the Morningstar Farms label, the products have an air of the breakfast table: a ham slice, a link sausage, a sausage patty, and a bacon strip. Miles first brought out these products in 1975, but the history of soy-based imitation meat in the United States goes back to the late 1920's, when T. A. Van Gundy developed imitation meat products based on soy which he sold through health food stores. In 1939, Worthington Foods of Worthington, Ohio, a firm founded by the vegetarian Seventh-Day Adventists, began manufacturing soybean-based meat on a large scale. Obtainable only through health food stores, as they still are today, the Worthington products included some sixty items ranging from imitation turkey to a fake scallop. I've tried a couple of products, which were attractive-looking and reasonably chewy, with a texture much like that of a thick slice of lunch meat.

When Worthington first started making meatless "meat," it depended on products made by putting soy flour through a machine known as an extruder, an old standby in the food processing business that is used for everything from corn curls to dog food. As the extruder pushes a flour-water mixture out of a small opening under pressure and heat, the water is flashed off as steam, making the product expand. The result is a mass with a firm, chewy texture something like that of meat. Add flavors, spices, and colors, and it looks and tastes rather meaty, too. The same process is used by other food companies making soy products destined for the human stomach, including Archer Daniels Midland. The latter has patented an extruded meatlike product under the name TVP (for textured vegetable protein). Extenders, for the most part, are extruded products.

In the 1950's, Worthington came up with a new soy product called "spun vegetable protein," which is produced by a process patented by Robert Boyer of the Ford Motor Company. Boyer got his idea from a similar process used in spinning rayon. Soy isolate is pumped through a spinnerette containing thousands of tiny holes and then into an acid bath that causes the mass to coagulate into threadlike fibers. The fibers are bland and white, but they have a meaty texture and are so versatile they can be made into almost any form of meat from a slice to a whole roast. Colors, flavors, and spices are added to achieve a meatlike taste. The spun product is more expensive to produce than the extruded product, making it more suitable for relatively high-priced meat analogs that completely replace meat, rather than for extenders. Most of Worthington's products are now based on spun vegetable protein.

After Miles acquired Worthington in 1970, the parent firm began planning a small line of spun vegetable products. Meanwhile, some of the big food companies tested similar products on the market. Depending on where you live, you may have found General Foods' Lean Strips, an imitation bacon, or several meatlike products manufactured by General Mills under the name Country Cuts in your supermarket. Both products were discontinued in 1976, although General Mills still markets its soy-based BacOs, imitation bacon bits in a jar.

In 1975, Miles finally began selling its meat analogs in super-markets all over the country. A 1977 article in *The Wall Street Journal* reports that Morningstar Farms products lost money in both 1975 and 1976. An improved version of the analogs was being prepared by late 1977, about the time the products disappeared from the freezer case in my supermarket.

Do the Morningstar products taste like meat? I first ate them as hors d'oeuvres at a press party when Miles introduced them, and the press—mostly food editors—was generally kind; some, however, mentioned the spices disapprovingly. I like spicy food, so I liked the products, which have a com-mendably meatlike texture. Since then, I've bought Morning-star products a number of times and found that they are easy to make (just put oil in a skillet and brown) and have prac-tically no shrinkage. What you buy is what you eat. One product, the breakfast strip, however, is so crisp it is prac-tically impossible to pick up on a fork. The protein content ranges from 5 grams per serving for breakfast strip, which has more egg albumen than soybeans, to a very impressive 12 grams for the sausage patty. A meat-based sausage patty of the same weight, according to Miles, has only 11 grams of protein.

The biggest problem with the Morningstar products is cost. Breakfast strips, for instance, cost 99 cents at my supermarket for a 5.25-ounce package, which means that a pound would cost about $3.00. Real bacon sells for less than $2.00 a pound. Bacon shrinks, but $3.00 is still a lot to pay for a pound of imitation bacon. The high price is due partly to expensive processing, partly to the high price of soybeans. At this writ-ing, soybeans are selling for $9.00 per bushel on the Chicago Board of Trade, less than the peak they reached during 1973 but more than twice as much as they sold for in 1972. The general trend is up. One way Miles is trying to get around the high price of their product is by advertising its benefits. Just above the name of the product on the Breakfast Links, for instance, are the words: "Cholesterol Free." To make sure you get the point, "NO ANIMAL FAT" is printed in capital letters on another area of the package. The protein content is played down.

Are there enough people in the United States who are worried about cholesterol but who want to eat meat with their breakfast to keep meat analogs on the market? Check your supermarket freezer case in a few years and see.

A few other products that use ample amounts of soy are also available in the supermarket, including Quaker Oats' Life cereal, General Foods' Fortified Oat Flakes (both are delicious cereals I'm adding to my shopping list), Gerber's Gerber High Protein (a baby cereal), and several brands of salted soybeans.

Soybean-based products, however, are more widely eaten in the United States than supermarket shelves indicate. Most of the soy produced for human foods in the United States goes into products made available by what is known as the institutional market—schools, nursing homes, hospitals, prisons, employee cafeterias, and the like. Institutions use large amounts of soy-based products because they are looking for more nutrition at lower price. That means that the soy-based foods institutions use are not the expensive analogs but the cheaper extruder-processed foods, many of them manufactured solely for the institutional market. Archer Daniels Midland, for example, sells most of its human soy products to the institutional market. These products end up in foods like pizza, chili, meatballs, and sloppy Joes, where subtleties like a beany flavor or a texture that falls short of meat are drowned in a sea of tomato sauce. The institutional market got its biggest impetus in 1971, when the federal government announced that up to 30 percent of the protein in the federal school lunch program could be vegetable protein. An estimated forty million pounds of vegetable protein are used each year in the program.

Nutritional studies indicate that vegetable protein does a good job of providing lots of protein for little cost. A research project conducted with young children in Guatemala by Dr. Ricardo Bressani of the Institute of Nutrition of Central America and Panama (INCAP) shows practically no difference in the quality of the protein between spun soy foods and milk. With both foods, the youngsters' nitrogen balance, the best indicator of protein value to the body, was essentially the same. The researchers concluded that textured protein had about 80 percent of the protein quality of milk. Dogs that ate the same textured protein food over a much longer period of time grew just as well as they did on a beef diet. Both the children and the dogs liked the food and neither man nor beast showed any adverse effects—including gas. Spun soy protein, it will be remembered, is made from soy isolates, which contain practically none of the sugars believed to be responsible for gas.

In another study, this one conducted by Suzanne Davis Khoury and Dr. Robert E. Hodges of the University of Iowa Hospitals, volunteers in a U.S. prison maintained "excellent clinical health" on a diet in which they received all their protein from spun vegetable sources for six months. The prisoners, like the Guatemalan children, had no complaints, but another group in the Khoury-Hodges study did. The investigators fed the same spun vegetable proteins to a group of medical school students and their wives for what was supposed to be a six-month project. A choice of four vegetable entrees was available: chicken, beef, ham, and seafood. The group accepted the imitation meats and seafood with some enthusiasm when hospital dieticians prepared them, but their appetites waned as time passed. When the students had to prepare their vegetable protein meals at home, most of them quit the project even though, as the researchers pointed out, the meals were easy to make. Some complained the diet was "monotonous." The choosy American consumer, apparently, is not quite ready for fake chicken or imitation ham. Not three times a day, anyway.

Consumer-oriented soy products may be relatively scarce today, but many food experts believe we will be eating more of them in the future. Walter Wolf has prepared figures which indicate that vegetable protein will undergo the biggest rise in sales of all fabricated foods by 1980. Most of that protein, it seems likely, will be based on soybeans. The Cinderella bean, as nutrition expert Max Milner of the International Nutrition Policy and Planning Program, Massachusetts Institute of Technology, points out, is readily available and we know how to process it cheaply, advantages no other source of vegetable protein enjoys. To get ready for the expected demand, researchers at the USDA's Northern Regional Research Laboratory in Peoria are creating new ways to use significant amounts of soy in food. The major effort is aimed not at putting new foods on our tables, but at putting more soy in traditional foods. On the face of it, fortifying foods we are accustomed to is a good idea, since it eliminates expensive efforts to convince the consumer to buy new products.

Unfortunately, the soy that does so many marvelous functional things in small amounts often has deleterious effects in larger quantities. Take bread, for instance. Put too much soy flour in bread, a prime candidate for soy fortification, and the loaf sags like the sales chart for extended hamburger. The same effect occurs with cakes and muffins. The reason why soy flour and other soy products have this depressing effect in baked products is that they lack gluten, the substance in wheat flour that helps bread rise. Soy products are more like powders than conventional flours. But there is a way around this problem—several ways, in fact.

"This is an ordinary loaf of bread baked with 22 percent soy isolate," said Donald Christensen, a research chemist at the NRRL, showing me a slide of a sorry-looking loaf that would make any baker weep. "And this," he said triumphantly, "is a loaf of bread baked with 22 percent soy isolate, wheat starch, and xanthan gum. A slice is as rich in protein as a piece of lunch meat." The loaf shown on the slide was high and mighty, just the way bread should be to please American tastes. Xanthan gum, a water-thickening agent discovered some years ago at the NRRL, now turns up in all sorts of processed foods (check the label on your salad dressing, a product in which xanthan gum is widely used). In breads, it functions as a gluten substitute, giving doughs and batters the strength and elasticity to rise with the carbon dioxide generated by yeast or leavening agents. Christensen ran through some other slides, all showing the kind of puffy loaves we love to eat, all made from mixtures of xanthan gum, unconventional high-protein flours, and starch. Corn germ bread. Peanut flour bread. Triticale bread. Cottonseed bread. Soy flour bread with various amounts of soy up to 22 percent.

"We even made a bread with nothing but starch and xanthan gum," put in Dr. Harold W. Gardner, the project leader on the xanthan gum work. "We call it 'empty calorie' bread. These breads are not made like bread; they are made like cake. But they fooled a taste panel. What we are doing is simulating the functional properties of bread."

The new process, according to Christensen, is actually simpler than ordinary bread baking, since it removes four steps from baking that are required with the use of gluten—no kneading, for instance. The process also works for cakes, muffins, quick breads, and doughnuts. "We made a cornstarch cake last week for a retirement party," he said. "They loved it." The use of xanthan gum in bakery products presents one problem, however. It's expensive. Although only about 2 percent of the loaf is xanthan gum, that ingredient boosts the price of a loaf to that of a specialty bread. Specialty breads do find a market in the wealthy United States, but the NRRL would like to see soy in the standard loaf of bread, not only here but in other countries.

Another way to produce an acceptable loaf of bread with a high soy content has been developed by Karl F. Finney, a cereal chemist, and Yeshajahu Pomeranz, director of the U.S. Grain Marketing Center in Manhattan, Kansas. The secret of the loaf is a sugar-free formula that incorporates small amounts of glycolipids, which are complexes of carbohydrates and fats, or sucrose palmitate instead of shortening. The latter is now approved for food use in the United States. The new formula allows 10 percent of the wheat to be replaced by 10 percent soy flour and 4 percent soy grits (coarsely ground soy flour), raising the protein content of the loaf significantly. A version of this formula which incorporates Vitamin C and malted flour, together with smaller amounts of soy flour, can be prepared by the home baker.

Soy also can be used to fortify other cereal-based food staples such as Mexico's tortilla, India's chapatti, the American South's cornmeal bread, Battle Creek-style breakfast cereals, and one of our nation's foremost contributions to the table, the snack. The much-maligned snack is one of the prime targets for soy fortification at the NRRL. Recently, chemical engineers Howard F. Conway and Roy A. Anderson turned out yards of crisp, puffy corn-soy curls that looked—and tasted—much like the corn curls beloved of U.S. television watchers, but which have up to 30 percent protein. The taste of soy just starts to come through when the curls are 40 percent soy flakes, indicating that the objectionable soy flavor can be masked in some snacks.

Will the lowly snack food be in the vanguard of soy fortification? I can hear it now: "Johnny, finish your bag of soy-corn curls; they're good for you."

3

SOYBEANS, ORIENTAL STYLE

Made of soybeans,
Square, cleanly cut,
And soft.

—INGEN,
Chinese Zen master

A PLATE OF WHAT LOOKED LIKE A NEW SNACK FOOD LAY ON the desk of Dr. C. W. Hesseltine at the U.S. Department of Agriculture's Northern Regional Research Laboratory in Peoria. It consisted of thin strips about four inches long and an inch wide with a crispy-looking coating. "Tempeh," I said, remembering the name of a similar-looking product I had seen in photographs in publications Hesseltine had sent me. "Try one," invited Hesseltine, who is chief of the Center's Fermentation Laboratory. I ate a strip and automatically reached for another. Good—in fact, very good. Crunchy, the way it looked, but with an unusual and intriguing taste. I kept reaching for more strips, along with Hesseltine and one of his associates, Dr. Hwa-Li Wang, a research chemist who was born in China.

"The only trouble with tempeh is that you can't stop eating it," said Hesseltine, a short man with an impish smile.

"It doesn't taste like soybeans at all."

"That's what everyone says. We tell them tempeh is made from fermented soybeans and they say: 'You're kidding.' I like it even better sliced thin and fried with a little bit of shrimp sauce. My family is always asking me to bring more tempeh home. It's the most acceptable soybean food to most Americans."

Hesseltine is the leading expert in the United States on fermented soybean foods, which are extremely popular in some Asian countries but almost completely unknown here. There is one exception: soy sauce. The dark, salty fluid known to all lovers of Oriental food is made by means of a fermentation process in which certain microbes (small, one-celled organisms) convert soybeans into a flavorful product that will keep indefinitely. Soy sauce, however, is just one of a number of fermented soybean foods which form an important part of the diet in some Asian countries. In Japan, soy sauce and miso, a paste that looks and spreads like peanut butter, are the chief means of providing food flavors. The Chinese like soy sauce and miso, but they also have a fermented soy treat peculiar to their own country: soy cheese, or sufu. Tempeh, one of the few fermented soybean products used as a main dish, is a common entree in Indonesia.

Hesseltine became interested in fermented soy foods in the 1950's, when the United States began supplying Japan with large amounts of soybeans. The new source of soybeans gave rise to complaints. In making miso and soy sauce, Japanese food processors found that the American varieties of soybeans didn't behave quite like their Japanese counterparts during fermentation, producing an inferior product. Hesseltine was consulted by American soybean growers, who were anxious not to lose their new market. He determined that some American soybean varieties were more suitable than others for fermented products and that small changes in processing would make the American beans usable in miso and other Japanese fermented foods. Thanks to his research, the marriage of American soybeans and Japanese fermented foods continues.

"The Japanese gave me this plaque," said Hesseltine, walking into his outer office. The wall plaque, in Oriental characters, is provided with a translation on the back: "By you,

Dr. Hesseltine, the righteous proofs of Japanese miso and shoyu [soy sauce] were established."

In the course of his work on Japanese fermented soybean foods, Hesseltine explained, he became interested in other Oriental soybean fermented products and, eventually, in a few nonsoybean fermented foods as well. By now, he has published a number of papers on the subject, some with Wang. The two have worked out efficient processes for making several of the traditional Oriental fermented foods on a commercial basis in this country, including tempeh, sufu, and a fermented version of soybean "milk." Fermented soybean foods, they believe, offer certain advantages over other soybean foods for this country. One major advantage is that they appear to cause little flatulence. Hesseltine and Wang sent samples of their tempeh to Dr. Doris Calloway of the University of California at Berkeley for use in tests she conducted on the gas-producing proclivities of various bean products. Tempeh, she found, was "essentially nonflatulent" when fed to human subjects. The exact reason remains obscure, but Wang suggests one reason might be the presence of the microbes, which produce an antibiotic effect that kills off gas-producing bacteria. Another reason may lie in the fact that in Asia, soybeans are customarily soaked before use in both fermented and unfermented foods.

"The water soaking," speculated Wang, "may take away the sugars."

A second big advantage enjoyed by fermented soybean foods is their taste. Unfermented soybean foods taste both bland and beany to most Americans but fermented foods have a strong but pleasant flavor that replaces the beany taste.

The best candidate for the fermented soybean product most likely to succeed here? It's tempeh, Hesseltine and Wang agree. Tempeh, Hesseltine pointed out, has a few added advantages over other popular fermented soybean foods: It not only tastes good and lacks flatulence, but it smells good and looks good, too. And it's easy to prepare—even in the home. Hesseltine describes its preparation in a typical Indonesian home in a number of his scientific papers. After the beans are soaked, they are placed in a basket in a stream, where the beans are trodden by foot until the hulls loosen and float away. Then the dehulled beans are boiled for a half hour. When they cool, they are wrapped in banana leaves with a bit of tempeh from a former batch, put in a corner of the room, and left alone for several days. At the end of that time, the microbe *Rhizopus oligosporus*, which was present in the old batch of tempeh, has grown over and around the beans, forming them into a solid cake with a whitish surface. You can eat it that way, fry it, or use it in soups and stews. The protein content is a very respectable 25 percent, about the same as meat.

Westerners apparently find tempeh easier to digest than some other soybean products. In *Legumes in Human Nutrition*, Dr. W. R. Aykroyd, former director of the Nutrition Division of the Food and Agricultural Organization of the United Nations, and Joyce Doughty relate the story of a group of European prisoners of war in a Japanese prison camp in the Philippines during World War II. The men found the boiled soybeans they were given hard to digest, so several Dutch prisoners from Indonesia, who were familiar with tempeh, suggested fermentation. After obtaining a microbial culture, the men soaked the beans and fermented them outdoors in the hot climate. The resulting product was "successfully used in treating protein and vitamin deficiencies" according to a British medical report cited by the authors. Thus far, the reason for the increased digestibility of tempeh isn't known, but most scientists assume it has something to do with the partial digestion of the soybeans by the microbes in the fermentation process.

The procedure Hesseltine and Wang worked out for making tempeh in this country has a number of up-to-date touches, including a pure culture of *Rhizopus* from the Fermentation Laboratory's vast collection of microbes, plastic bags instead of banana leaves (an item in short supply in Illinois), and an incubator. To see modern tempeh making in progress, the two researchers took me to a laboratory manned by technician

Earl Swain, who had arranged a small exhibition for my benefit. Walking down a long corridor outside Hesseltine's office, we passed laboratory after laboratory in a seemingly endless procession. Fermented soybean foods, explained Hesseltine, are really a sideline in the Fermentation Laboratory, the largest of the four laboratories at the NRRL. Its major work revolves around the microbial conversion of agricultural commodities into foods, feeds, drugs, and other products widely used in the United States. To that end, the Laboratory maintains a world-famous collection of microbes that includes many thousands of bacteria, molds, and yeasts. "We have the world's largest collection of industrial microorganisms," said Hesseltine.

When we arrived at our destination, Earl Swain opened a refrigerator and took out a small glass dish that contained a solid, cakelike mass covered with a fuzzy, whitish substance. "This is fermented tempeh," he said. "We started it yesterday. And this is what it looks like before it's inoculated." He took a vial and another glass dish out of the refrigerator, the latter full of yellowish beans. "To make tempeh, we inoculate the chopped and soaked beans with a little of this," he said, indicating the vial, which contained a fuzzy white sheet of *Rhizopus* floating on a yellow liquid. "We use one teaspoon per pound of soybeans." The inoculated product, he said, then goes into a small commercial incubator.

I sniffed the fresh tempeh. A pleasant smell.

The fresh product, the Peoria researchers found, can be readily frozen and thawed and the thawed product tastes just as good as the fresh one. The NRRL method of making tempeh has been publicized in a number of papers and more and more aspiring tempeh makers are asking the NRRL for *Rhizopus* cultures. In the first six months of 1976, Swain said, he sent out two hundred samples of tempeh inoculum. As of this writing, the NRRL was still sending out inoculum but the practice may come to a halt. "We can hardly handle the requests coming in," said Hesseltine. Among those who have requested *Rhizopus* are The Farm, a large vegetarian commune in Summertown, Tennessee, and the Indonesian Tempeh Company of Unadilla, Nebraska, the only commercial firm selling tempeh in this country. Its customers are grocery and health food stores in Nebraska. The Farm, the largest commune in the United States, has about 1,000 members who eat no meat or dairy products. Tempeh is an important protein source in their diet.

"The last I heard, The Farm was making sixty pounds of tempeh a day," said Hesseltine.

On a laboratory table next to the tempeh was a brownish, pastelike substance that looked much like peanut butter. "Try some miso as long as you're here," said Hesseltine, indicating the brownish paste. "It's the most popular soybean product in Japan after soybean sauce." I took a dab with a spoon. It was salty and sharp, with an appealing, cheeselike taste. The Japanese, according to Hesseltine, do not eat miso directly but use it as the basis for soups and sauces. Every morning, most Japanese eat a soup made with miso stock. Hesseltine himself enjoys miso spread on a cucumber slice, particularly the strong red miso.

"We contributed something to the Japanese miso industry," he said. "They used to use mixed cultures but our work convinced them to use pure cultures for the second phase of fermentation."

Miso, as Hesseltine explains in his papers, is a much more difficult food to produce than tempeh, making miso more suitable to a commercial enterprise than to the home kitchen. Soy sauce, too, requires a complicated series of steps. Since neither is particularly high in protein (miso has 10 percent, soy sauce, 5 to 6 percent) and both are used primarily for flavoring, the protein they furnish is probably not a significant part of the Asian diet. Two other popular fermented soybean foods used to add flavor are hamanatto, which is used in China, Japan, the East Indies, and the Philippines, and natto, which is used primarily in Japan. The latter has a dark, slimy appearance and a peculiar smell. Even Hesseltine, who likes almost all fermented foods, doesn't like natto; few Westerners do.

"It doesn't look good and it smells terrible," he said. The test of good natto, he noted, is sliminess. If you hold up two beans and a string of slime runs between them, the natto is considered suitable for use.

Besides updating traditional soybean fermented foods, Hesseltine and Wang develop new ones. "Would you like to try our latest soybean fermented food?" asked Wang.

I was enthusiastic, so she went out of the room and returned with a small cup. "Soybean yogurt," she explained. I spooned some into my mouth. I like dairy yogurt but this was better—unflavored, which gave it the slightly sour taste of yogurt, but with a delicate taste of its own that did not resemble soybeans at all. The texture was soft and creamy. Wang explained that the new product is made from soy milk, a traditional Asian beverage made from soybeans and water, but the milk is a special blend made specifically for yogurt. "We are patenting the yogurt because it has some unique properties," she said. "For one thing, it needs different bacteria than cow's milk yogurt. One of the bacteria actually eats sugars." The product produces little flatulence and the protein content can be manipulated by raising or lowering the amount of water, unlike dairy yogurt. The soy-based yogurt offers so many advantages I have no trouble envisioning it as a product in the supermarket dairy case, right next to dairy yogurt.

After tempeh and yogurt, the fermented soybean food with the greatest potential for adding a sizable amount of protein to the American diet in an appealing form is probably sufu, a fermented form of the ubiquitous bean cake or tofu that turns up in so many Asian dishes. Sufu is the closest thing there is to a traditional soybean cheese. The NRRL had none on hand when I was there, but before my Peoria trip I had visited the largest market in New York City's Chinatown and, after much questioning of various employees, purchased a jar of sufu. The confusion arose because in Chinese markets, sufu is not called sufu but "bean curd" or "bean cake." As soon as I saw sufu, I realized it has an image problem. Sufu looks unappetizing—not as unappetizing as natto, but distinctly unpleasant. Picture grayish chunks of some odd-looking material floating in a murky fluid, like biology specimens in a bottle, and you have a typical bottle of sufu.

Sufu looks so bad that my husband, who has faithfully eaten a number of odd-looking sources of protein I have purchased over the years, refused it. It took a little courage for me to tackle one of the grayish lumps myself but I finally ate one. To my surprise, it was good, rather like a tangy dairy cheese but with a distinctive, nonbeany flavor of its own. For wider appeal in this country, though, sufu definitely needs a new image. The problem is that its preparation does not really lend itself to a different mode of packaging. Sufu is made by inoculating cubes of tofu, the bland Asian soybean cake, with a fungus and then placing the cubes in a pickling fluid composed of rice wine and brine. Since most of the flavor is acquired during the pickling stage, sufu must be immersed in the fluid. Considering that sufu has 10 percent protein and is low in cholesterol, however, it might be worthwhile for the American food processing industry to devise a method of making sufu that produces a product more appealing to Western tastes.

Soybeans are the major source of fermented foods in Asia but fermentation can also be used to produce nutritious and good-tasting food products from other raw materials. In Indonesia, two cheap forms of tempeh, *ontjam* and *bongkrek*, are made from peanut and coconut press cake, the residue left after the oil is removed. According to Dr. Keith H. Steinkraus of Cornell University's Agricultural Experiment Station, poor Indonesians eat more of these forms of tempeh than soybean tempeh. *Tape*, an alcoholic (6 percent) fermented Indonesian food, has a rice or cassava base. The protein content of the rice-based *tape* is 16 percent, that of the cassava *tape* 6 to 8 percent. White rice has about 7 percent protein, cassava only 1 to 2 percent protein, making *tape* much more nutritious

than either of the foods it is made from. *Idli*, a widely used
Indian food product, is based on a fermented mixture of rice
and black grams, a variety of bean. When fried, it is eaten like
a pancake.

Steinkraus, who has experimented with all these foods,
thinks they may someday play a role in our diet. "As our population continues to increase and supplies of meat and other
animal products become relatively more expensive, America
and the Western world may find it desirable to adapt more of
the exotic Asian foods to their diets," he writes in a recent
issue of *Food and Life Sciences Quarterly*, a Cornell University publication.

If fermented Asian foods ever reach the U.S. consumer,
though, Hesseltine believes the impetus will come from the
Asian food industry, not our own. We produce a few fermented foods, he acknowledges (including cheese, beer, and
sauerkraut), but Asians have much more familiarity with fermentation, thanks to their long interest in fermented soybean
foods. The Japanese fermentation technology is the most sophisticated in the world. "Our food industry just isn't geared
to using fermentation," Hesseltine told me. "I have a feeling
that if the marketing effort is put forth to introduce a new fermented food like tempeh, it will be done by a Japanese firm
in the United States." A Japanese firm is already making Kik-
koman soy sauce in a plant in Wisconsin, he noted. In addition, several small firms operated by U.S. citizens of Asian
descent produce miso and sufu in the western United States.

It's understandable that Asians know much more about
using soybeans as a human food than we do; they've had the
bean through most of their long recorded history. The soybean originated, most agronomists agree, somewhere in northern China, probably Manchuria, where the heavy early-summer
rainfall suits the growing plant. A Chinese tradition says the
emperor Sheng-nung first described the soybean in 2838 B.C.,
but the first reliable historical references to the cultivated
bean are dated several thousand years later. A Chinese ruler,
Lui An, is credited with creating tofu back in the second cen-

tury B.C. Tofu is not a fermented product, although it's often called "Chinese cheese," but a solidified cake made from an extract of soybeans and water. Between the sixth and eighth centuries A.D., when Buddhism was spreading throughout Asia, Buddhist monks took soybeans from China to Japan. Early Japanese tax documents dating from the eighth century mention miso and soy sauce. Tofu was probably first prepared in Japan by Buddhist monks as part of their vegetarian diet. When the monks opened vegetarian restaurants in their temples in the twelfth century, Japanese laymen ate their first tofu. Apparently it was love at first bite. Since that time, references to the bean cake turn up often in Japanese records and literature.

PROTEIN CONTENT OF TOFU

TYPE OF TOFU	PROTEIN CONTENT (% PER 100 GM)
Tofu	7.8
Doufu (Chinese-style firm tofu)	10.6
Kinugoshi	5.5
Thick Agé	10.1
Agé	18.6
Soy milk	2.0
Ganmo	15.4
Grilled tofu	8.8
Dried-frozen tofu	53.4
Dried Yuba	52.3
Pressed tofu and savory tofu	22.0
Dry soybeans	34.3
Defatted soybean meal	49.0

Note: Regular Japanese tofu varies in protein content from 6 to 8.4, kinugoshi and rich soy milk prepared at tofu shops from 4.9 to 6.3, commercially distributed soy milk from 3.6 to 5.8. Differences in composition depend primarily on the method of preparation, the type of soiidifier, and the grade and protein content of the soybeans used.

Source: *The Book of Tofu*, 1975.

A well-known proverb was created in the seventeenth century by the Chinese Zen master Ingen in praise of tofu. It goes:

> Made of soybeans
> Square, cleanly cut,
> And soft.

Tofu, soy sauce, and miso are still the big three among soybean foods in China and Japan. Soy milk, a nonfermented beverage that is the starting point for tofu, is also popular. In Japan, soy milk is marketed in special shops devoted to soy products, in supermarkets, and even on the street. Soy milk is fairly simple to make from scratch if you have a food grinder or blender. You grind or blend soybeans, add water, cook the mixture briefly, then pour it into a bag made of loosely woven cloth, such as cheesecloth. Squeeze out the fluid, cook briefly again, and it's ready to drink. The percentage of protein in soy milk varies according to its water content but several formulas offer about the same protein content as milk—around 3.3 percent—or even more. Like soybeans, though, soybean milk is low in methionine, one of the essential amino acids. Soy infant formulas are usually fortified with methionine and calcium, in which soy milk is also low.

To make tofu from soy milk, you add a *solidifier* (seawater is the ancient *solidifier*) and press the thickened curds into a flat box. After some of the fluid is pressed out, tofu can be sliced like cake. Unlike fermented soybean products, which are highly flavored, tofu has practically no taste, at least to the average Westerner. Flavor is acquired by putting cubes of tofu in stews or soups flavored with fermented soybean products. The protein content of ordinary tofu varies from about 8 to 10 percent, although a few specialized types of tofu have considerably more protein. Dr. Doris Calloway of the University of California at Berkeley found that the flatulence-producing capacity of tofu was practically nil, even after subjects ate 300 grams (10.5 ounces). That amount supplies as much protein as 70 grams (2.5 ounces) of soybeans.

The Japanese obtain much of their protein from soybean products like these. In *Legumes in Human Nutrition*, Aykroyd and Doughty estimate the annual production and importation of soybeans in Japan at a million tons, more than half of which is used to make products like miso, soy sauce, and tofu. The average consumption of soybeans per person per day is estimated at 64 grams (2.2 ounces). No figures on soybean consumption are available for China but it's estimated that the Chinese eat about 18 grams (.6 ounce) of soybeans per person daily. Few soybeans are eaten directly in Japan, the way health food enthusiasts in this country eat them. Almost all are processed, either in the home or in small shops or factories. The fact that the soybean is primarily a processed product makes it difficult to export to other countries, point out Aykroyd and Doughty, because countries unfamiliar with soybeans lack the processing technology.

A new book written by an American and a Japanese gives an idea of the importance of one processed soybean product, tofu, in Japan. According to *The Book of Tofu* by William Shurtleff and Akiko Aoyagi, tofu making is both an art and a business in Japan. Tofu is now made by large manufacturers and sold in the Japanese version of the supermarket, but it is also made by small businessmen in "tofu shops," often according to traditional methods handed down through the centuries. There are some thirty-eight thousand tofu shops in Japan. Shurtleff and Aoyagi describe the operation of a typical shop:

> After stirring nigari [a solidifier] into the soymilk in his large cedar curding barrel, the tofu craftsman covers the barrel with a wooden lid and allows the nigari to begin its work. Slowly it solidifies the soybean protein, which forms into curds and separates from the whey. After 15 or 20 minutes, the tofu maker rinses off a large, handsome bamboo colander and wraps its underside with cloth. He sets this on the surface of the mixture in the barrel and it slowly fills with whey (the cloth keeps out

the finer particles of curd). The whey in the colander is ladled off and reserved for later use, and the colander is then weighted with a brick and replaced until it is again full. This whey, too, is ladled off into a large wooden bucket where it forms a billowy head of foam. When all the whey has been removed, only white curds remain in the barrel.

It is these curds, the authors explain, that are ladled into cloth-lined boxes and pressed to make tofu.

Until I read *The Book of Tofu*, I thought the only form tofu came in was the fresh white cubes I had eaten in Chinese and Japanese restaurants, but according to Shurtleff and Aoyagi, there are seven different forms available in Japan and many more in China. The white cubed product I was familiar with is the most common one in both China and Japan. Next in popularity in Japan is a deep-fried product known as agé, which comes in at least three varieties. Other popular Japanese tofu products are grilled tofu, a cold-weather item, and frozen tofu, which is not, as its name suggests, an offshoot of the space age but an ancient product that used to be made by putting slabs of fresh tofu in snowdrifts (today the Japanese use freezers). Kinugoshi, a particularly soft version of tofu— so soft that one variety cannot be cubed—is available in both China and Japan. The Chinese have a number of tofu specialties of their own, including pressed tofu, a firm product that has much less water than ordinary tofu and about 22 percent protein. They also have a dark brown, salt-dried product with a cheeselike consistency and, of course, sufu.

A soybean product that is allied to tofu but contains much more protein is yuba, which has 52.4 percent protein. Sold in the form of dried sheets, yuba is really the "skin" that forms on hot soybean milk during processing. Yuba is popular in both China and Japan. In Japan, it is sold only in gourmet yuba shops, most of them in the Kyoto area, but in China it is much more widely available, being sold both in yuba shops and in grocery stores. In recent years, factory-processed yuba has become a big-selling item in China. The most unusual form of yuba is an imitation meat that is popular with Chinese vegetarians. To make it, fresh yuba is pressed into molds made in the shape of various animals or parts of animals, such as ducks, pigs' heads, and fish, and heated. When the animal-shaped product is taken out of the mold, it is flavored and served cold, fried, or simmered in a broth. The meatless product is sold under such traditional names as Buddha's Duck and Molded Pig's Head. When Shurtleff and Aoyagi visited Taiwan recently, they found restaurants specializing in vegetarian Buddhist cuisine offering a wide variety of yuba "meat."

Your chances of finding Buddha's Duck in your town are rather slim, but some traditional Oriental soybean foods are surprisingly available in the United States. Soybean sauce, of course, is everywhere but miso and tofu are available in any Oriental food store and even in many health food stores. Some Oriental food stores also carry jars of sufu. Probably the hardest product to obtain is tempeh. Fresh tempeh is very perishable and no frozen form of the product is available as yet, making it almost impossible to find tempeh unless you live in the area of Nebraska served by the Indonesian Tempeh Company. If you're really serious about trying tempeh, though, you can now make it in your own kitchen. The Farm, the Tennessee vegetarian commune, began selling a Complete Tempeh Kit including starter, in mid-1977. Tempeh starter, split soybeans for use in tempeh, and instructions on how to make the Indonesian dish are available separately. The Farm also offers two sizes of its Tofu Box Kit.

Along with literature for their kits, The Farm sent me a brochure including a photograph of a "tempeh barbecue" at the commune. As a young woman waits with a plate, a man flips squares of frying tempeh on a grill. My mouth watered as I remembered the tempeh I ate in Peoria.

Tofu is easier to make than tempeh and you don't even need The Farm's kit—just implements, a recipe, and a supply of solidifier. The latter may be magnesium chloride, calcium chloride, magnesium sulfate (Epsom salts), calcium sulfate (gypsum), or even vinegar or lemon juice. You'll probably like tempeh, but for many Westerners the first taste of tofu—

and even subsequent tastes—presents something of a culture shock. I've eaten tofu a number of times in Japanese restaurants in both the fresh and deep-fried agé form but it has never really appealed to me in traditional dishes. I find fresh tofu too bland, agé too beany. Some cold, rice-filled agé pockets I recently ate for lunch in a Japanese restaurant reminded me of wet cardboard in texture and taste. Frozen agé pockets I purchased in a Japanese food store and heated at home for hors d'oeuvres were better but didn't arouse my enthusiasm. But when I made a simple Western-style spread using tofu, peanut butter, honey, and lemon juice as described in *The Book of Tofu*, and ate it on a sandwich for lunch, it was delicious. I ate two sandwiches instead of one. I suspect that for the Westerner, the best way to enjoy tofu is to use the bland fresh form in some of the many Western recipes given in vegetarian cookbooks. In these dishes, its blandness becomes a positive asset, enabling it to mingle happily with everything from peanut butter to avocados.

Some U.S. researchers, in fact, have discovered that the very lack of taste in fresh tofu that bothers many Westerners makes it suitable for use as an inexpensive meat substitute or extender. Vivian Yeo and G. H. Willington of the Department of Animal Science at Cornell University and Keith Steinkraus recently put fresh tofu in that most American of all products, the hamburger. Even tofuburgers that had up to 75 percent tofu received surprisingly good acceptance from a taste panel, if the tofu used was firm (firmness can be increased by reducing the water content in tofu), was colored red, and was flavored with meat. The firm tofu, the researchers note, produced a hamburger that has about 6 percent *more* protein than ordinary hamburger, which is about 18 percent protein. Like soybean extenders already in use, tofu lessened cooking loss, a big plus in hamburgers and processed meats.

Another food research team, this one at the Food Science Department of the University of Florida, has made tofu's close relative yuba into an imitation meat product that may be adaptable to the American market. L. C. Wu and R. B.

Bates heated soy milk in flat steel pans, removed the resulting films, and piled them on top of each other. The mass was wrapped tightly in plastic film and heated. The heat and pressure made the sheets of soy film adhere together tightly, forming a thick slab. When chunks cut from the slabs were flavored with bouillon and browned in oil, they smelled, looked, and chewed like meat. Did they taste like meat? Not really, admit the researchers, but the lack of meatlike flavor is fairly easy to overcome with flavoring and spices. They see U.S.-style yuba chunks as an inexpensive and nutritious meat substitute in casseroles and salads and, possibly, even as a new product that can stand on its own, like pasta.

An Oriental soybean food which arouses considerably more interest among U.S. food researchers today is soy milk. This beverage, a popular drink in Japan, has intrigued scientists here for decades as a possible milk substitute. A few commercial soy milks are already available. I've tried powdered Soyagen, which is made by Loma Linda Foods, a large manufacturer of vegetarian foods, and found it very palatable. Soyagen includes added oil and calcium to bring the composition closer to that of milk, as well as sugar, Vitamin B_{12}, and Vitamin D. A much sweeter infant formula is also available. The major market for these drinks in the United States is the sizable proportion of U.S. children—from 7 to 15 percent—who are allergic to cow's milk. Some adults allergic to cow's milk also use soy milk, as do some vegetarians. Many vegetarians, however, prefer to make their own soy milk, starting from scratch with soybeans. The Farm produces some eighty gallons of soy milk a day in its own dairy.

Passionate vegetarians swear by soy milk, but nonvegetarians like myself who drink it find it somewhat inferior to cow's milk. The faint beany flavor lingers on; the texture or "mouthfeel" is a little grainy; and the color is creamy, not white like milk's. But some new soy milk formulas that are coming out of U.S. laboratories may overcome all these drawbacks. Dr. Gus C. Mustakas of the Northern Regional Research Laboratory has developed a soy milk formula known as LPC—for "lipid-protein concentrate"—that is creamy white and has a milklike "mouthfeel." The drink, which starts with full-fat soybean flour, incorporates several breakthroughs. A special milling process reduces particle size, making the drink very smooth and enabling it to remain in the refrigerator for up to a week without separating into sediment and liquid, a drawback found in most other soy milk formulas. (I have some freshly made Soyagen in my refrigerator right now that has separated into these two phases.) The addition of oil during processing increases smoothness and produces a white color almost like that of milk. In addition, the drink doesn't induce flatulence because sugars are removed during processing. It emerges from the process as a liquid, just like milk, but it can be dried to a powder form. The beverage, alas, is not commercially available yet.

Earlier, Mustakas developed another soy milk formula that also starts with full-fat soy flour but doesn't incorporate all the steps in the LPC formula. Although the resulting beverage isn't quite as milky as LPC, the small particle size, extra oil, and artificial cream flavor make it a smooth, good-tasting beverage that might find friends here. At present it is being manufactured only in Mexico. In general, soy milk finds a better market in developing countries, where dairy products are unavailable or too expensive for most people to buy. A few big U.S. firms are marketing soy beverages in developing countries, including the Coca-Cola Company. Their latest soy formula, which is built around flatulence-free soy isolate, is being test marketed in Mexico and Brazil, and another Coca-Cola soy milk formula was introduced into Brazilian schools in 1977. Thus far, the soft drink company hasn't test marketed their soy milk formulas here, but some large firm may well be preparing such a campaign right now.

While we're waiting for American firms to bring out their versions of Oriental soybean foods, though, I haven't quite given up on the more traditional soybean products. I'm ready to concede that even fresh tofu, which I find a little boring, has more oomph when purchased directly from a neighborhood tofu shop. And the neighborhood tofu shop, according to Shurtleff and Aoyagi, may be a reality someday in this country. They envision a time when small tofu shops will be as common here as bakeries, offering all the forms of tofu Oriental people know and love. And if the tofu shop succeeds on the American scene, why not the tempeh shop?

4

THE MUSICAL FRUIT

Beans, beans,
The musical fruit.
The more you eat,
The more you toot.

—CHILDREN'S RHYME

ALMOST EIGHT THOUSAND YEARS AGO, INDIANS IN WHAT IS today the Ancash Province of Peru cultivated the same kind of common beans and lima beans we eat today. Dr. Lawrence Kaplan, Chairman of the Department of Biology at the University of Massachusetts, excavated Guitarrero Cave in Ancash recently and found dark reddish-brown and dark red beans at the same level. Some of the beans look like today's kidney bean while others are more rounded, but both types are specimens of *Phaseolus vulgaris,* the common bean we eat today in the form of kidney beans, navy beans, and other well-known beans. Some of the Guitarrero Cave beans were dated by means of radiocarbon dating techniques at 7680 B.P. (before present), plus or minus 280 years, which makes the cache of Peruvian beans the oldest domesticated beans in this hemisphere and possibly in the world. Some beans from the site may be even older. Another group of beans Kaplan found at the same level as the eight-thousand-year-old beans are clearly lima beans. The domestic status of both groups of beans is "beyond doubt," according to Kaplan. Wild beans that have been found in Central and South America, he explains, are smaller and have a distinctive mottled appearance, among other characteristics. The Guitarrero beans are as large as more modern beans and have a similar coloration.

"The people of Guitarrero Cave practiced cultivation of common and lima beans between 5500 and 8500 B.C.," Kaplan says.

Earlier, Kaplan, who is the leading authority in the United States on ancient beans, found other domesticated beans that date far back into prehistoric times. Beans he excavated in Coxcatlan Cave in the Tehuacán Valley in Mexico have been radiocarbon dated at seven thousand years and beans he found in another Mexican site, the Ocampo caves, at six thousand years. Ocampo also yielded the oldest wild beans, which are from nine thousand to eleven thousand years old and belong to the same family as the decorative plant known in this country as the Scarlet Runner.

The evidence seems clear: Man has been cultivating and eating beans for a very long time, and with good reason. Dry beans have from 20 to 25 percent protein, on the average, with a few rising over the 30 percent mark. Soybeans have 34 to 37 percent protein, the highest among the beans with

PROTEIN CONTENT OF SELECTED LEGUMES

CROP	AVERAGE PROTEIN CONTENT (%)
Peanuts	25.6
Soybeans	38.0
Phaseolus vulgaris (Most U.S. dry beans)	22.1
Dry peas	22.5
Chick peas	20.1
Broad beans	23.4
Lentils	24.2

Source: FAO, 1969 and 1970.

the exception of a few noncommercial beans such as the winged bean (see Chapter 8). Peanuts, which are really a legume, have 25 percent protein. This high protein content makes beans the best source of protein among the plants. If you're a strict vegetarian (no meat or dairy products) by necessity or choice, you must depend largely on beans to supply your protein. Millions of people in the world *do* depend on beans to supply all or most of their protein. According to Dr. David Pimentel of Cornell University, about 20 percent of the protein currently available to man comes from beans. In many developing countries, an estimated 50 percent of the population gets nearly all its protein from beans.

All the beans and peas, as well as peanuts and lentils, belong to the big botanical family Leguminosae, members of which are usually referred to as legumes. It includes about thirteen thousand species and six hundred genera, only about a dozen of the latter being commercially important. Most of the dry beans we eat in this country belong to a single species known as *Phaseolus vulgaris*. Our most popular dry beans, the navy, pinto, and kidney beans, are all members of *Phaseolus vulgaris*. So are black beans, Great Northerns, red beans, and small whites. The blackeye pea, the lentil, the pea, the garbanzo, the peanut, and the soybean, on the other hand, belong to different genera. Altogether, we in the U.S. eat about eighteen to twenty different kinds of peas and beans, a fairly wide variety. Michigan is our leading bean-growing state, partly because of its relatively dry summer climate, an important factor in bean growing. Ninety-five percent of the navy beans grown in the United States are grown in Michigan.

Americans (including, until recently, this American) often think of beans as fresh "green beans," which are one of our most popular vegetable side dishes. This member of *Phaseolus vulgaris* falls short on protein because we eat it not in a mature dry stage, but in an immature fresh stage in which the seeds are very small. Fresh green beans have only about 10 to 15 percent protein, dry weight. (Call green beans "snap beans" if you're talking to an agricultural scientist or a farmer; "green beans," to them, means the immature stage of any bean.) Bean sprouts, on the other hand, have somewhat *more* protein than seeds because of the loss of low-protein components in the sprout and the effects of protein synthesis in the germinating seed. A recent study by Anne M. Kyler and Rolland M. McCready of Colorado State University and the U.S. Department of Agriculture, respectively, shows that the sprouts of alfalfa, lentils, mung beans, and soybeans all contain more protein than the seed when the contents are calculated on a dry-weight basis (seeds or sprouts minus water). Mung sprouts, the traditional Chinese bean sprout, showed the greatest increase in protein content over the seeds, about 19 percent. Since fresh sprouts are mostly water, though, you

MAJOR DRY BEANS GROWN
IN THE UNITED STATES

CLASS	METRIC TONS, 1974
Navy	314,900
Great Northern	96,500
Red kidney	66,600
California blackeye pea	49,500
Pink	46,700
Large lima	30,400
Small white	28,100
Baby lima	26,000
Pinto	22,400
Small red	20,200
Black turtle	11,200
Cranberry	8,300
Garbanzo	3,800
Flat small white	2,000
Other	15,700
	943,700

Source: USDA Statistical Reporting Service.

get only about 5 to 13 percent protein for each 100 grams (about 3.5 ounces) of sprouts.

Every country has its own legume favorites. In India, where beans and peas are called dal or dhal, the most popular beans are several kinds known as grams, the pigeon pea, the chick pea, and the lentil. The cowpea is the favorite in Africa, along with the peanut. Central America's most popular bean is the black bean, while kidney beans lead the list in Mexico. The soybean, as might be expected, is the number one bean in Japan, followed by the mung bean. In Europe, people eat lentils, blackeye peas, peas, and broad beans, among other species. Beans are often eaten whole, the way we eat them, but in various countries they are also served mashed, ground to a powder, fermented, or extracted into a drink or watery solid product like tofu.

Our own bean consumption in the U.S. is only about six pounds per person per year, a figure which has been slowly falling for many years. Beans are a much bigger food item in developing countries and even in some highly developed nations such as Japan. The Japanese eat about 64 grams (2.24 ounces) of beans per day, most of them in the form of processed soybean products. Brazilians eat about 68 grams (2.38 ounces) a day. Since the average-size bean weighs about a gram, that means sixty beans spread out over three meals a day. Indians eat about the same amount. Some people eat many more. In a world survey of bean-eating habits, one Indian family was recorded as eating more than 420 grams (14.7 ounces) of beans per day per person. The Chiga tribe of Uganda ate about 400 grams of beans per day per person. In most countries, there is a definite relationship between family income and bean eating. As income rises, people eat fewer beans. But in South America, everyone eats lots of beans—rich, poor, and middle class alike.

Why don't Americans eat more beans? Bean growers who would like to sell more beans, and nutritionists who would like to see Americans getting more of their protein from cheap and efficient plant protein sources, often ponder the question. Part of the answer seems to be that we are a rich people, comparatively speaking, and bean consumption, as indicated above, tends to fall with rising income. In a 1958 survey of bean eating in the United States, people with the lowest incomes in both urban and rural areas ate four times as many beans as people with the highest incomes. For the rich, or aspiring rich, meat becomes a sort of status symbol, attractive not only for its good taste and protein value, but for its ability to evoke envy and make its consumers feel as though they are enjoying the better things of life. Silly? Of course. But our steak restaurants, full of executives on expense accounts and people celebrating special events, are built on creating such psychic satisfactions.

Nevertheless, some respected observers of the American food scene are predicting an upturn in popularity right here in the rich United States for animal protein's closest competitor, the lowly bean. "I think the decline in bean consumption will be reversed because of the need for protein around the world," says Dr. M. Wayne Adams of Michigan State University, one of the nation's leading bean researchers.

Adams spells out his arguments for a bigger and better future for beans in a paper he wrote for the National Science Foundation with two other bean researchers, Dr. Louis B. Rockland of the U.S. Department of Agriculture's Western Regional Research Laboratory, and Dr. Gordon W. Monfort of the California Dry Bean Council. In an increasingly energy-short world, they point out, protein-rich dry beans are more efficient converters of energy to protein than almost any other plant. Soybeans have a 2.6 ratio of energy input to protein output, dry beans a 3.7 ratio. With the exception of corn (3.61), other plants lag far behind. Adams and his coauthors believe the more successful dry bean growers in the United States could reduce the present input-output ratio by one-half, making bean protein even more of a bargain. One economic factor in raising beans is that they make some of their own nitrogen fertilizer by means of biological nitrogen fixation (see Chapter 2). Thus, although beans raised in this country usually do receive some nitrogen fertilizer, they require less than most other plants.

Before beans can occupy a bigger place in the American diet, however, bean experts caution that a number of problems have to be overcome. The status symbol value of meat isn't the only reason for its desirability. In nutritional terms, the animal proteins—meat, fish, milk, and dairy products—are more complete sources of protein than the plant proteins. Beans fall short in the essential amino acid methionine and, to a lesser degree, in cystine. Both are often called the sulphur amino acids. Some beans are also a little short in tryptophan and isoleucine. On the other hand, beans are high in lysine, an amino acid in which all the widely used cereal grains, such as wheat, rice, and corn, are low.

The BV of most beans ranges from about 40 to 70, with an average of about 50. Again, soybeans score better, about 75. Meat has an average BV of 80, milk 83, and eggs 96. Even rice, the most balanced grain, scores about 70, better than any bean except soybeans.

Most of the nitrogen balance studies have been carried out with laboratory animals, usually rats, because certain studies require killing the animals to make measurements. But some studies have used human subjects. The procedure is to feed

BIOLOGICAL VALUE OF SOME LEGUMES

VARIETY	BV (%)
Cajanus cajan (pigeon pea)	46–74
Phaseolus vulgaris (black bean)	62–68
Cicer arietinum (chick pea)	52–78
Lens esculenta (lentil)	32–58
Phaseolus aureus (golden and green gram)	39–66
Phaseolus mungo (mung bean)	60–64
Pisum sativum (garden pea)	48–49

Source: Bressani, INCAP, 1972.

them protein foods, measure their nitrogen intake, and then measure the amount of nitrogen in their feces. The difference between the nitrogen consumed and the nitrogen excreted gives a measure of the nitrogen retained by the body. In a 1952 study described in the *Journal of Nutrition*, human adults had a fecal nitrogen reading of 14.5 percent of nitrogen intake when they ate eggs, but a fecal nitrogen reading of 20 to 21 percent of nitrogen intake when they ate lentils. Lentils, incidentally, are one of the more digestible beans. In other studies described in the same article, children who ate maize and black beans had a fecal nitrogen reading of 30 percent. It was only 18.1 percent when they drank milk.

With regard to beans, all these figures refer to *cooked* beans; raw beans contain proven antinutritional factors that are actually toxic in the case of some varieties. Rats fed a diet that included 10 percent raw navy beans in studies at Michigan State University died in a few weeks. Rats fed on a diet that included 20 percent raw navy beans died within ten days! Adding synthetic methionine to the raw beans reduced mortality, but the rodents still grew much more slowly than normal. The culprits in raw beans, scientists believe, include a component that cancels the action of a digestive enzyme, trypsin, and another component that leads to deleterious changes in the blood. Luckily for bean lovers, heating destroys toxic components in beans. There's evidence from other studies that not all raw beans are as toxic as navy beans, but don't take chances: Cook beans thoroughly to destroy toxins and improve nutrition. Besides, uncooked beans are hard on the teeth.

Do the relatively low PER and BV of beans mean that people who depend primarily on beans as a source of protein are doomed to malnutrition? Not at all. Millions of people throughout the world use beans as a major source of protein and thrive, including thousands of vegetarians in the United States. The secret is supplementation, the magic word in vegetable protein nutrition. Before I began reading about nutrition, I wondered why the Mexican "combination" dinners

I enjoyed at my favorite Mexican restaurant, Acapulco's in New York City, included both rice and kidney beans which, to me, spelled a double dose of starch. I no longer wonder. Beans and rice do have starch, of course, but the mixture of rice, a grain, with kidney beans also offers the diner a nutritionally balanced source of protein. All the popular food grains are low in lysine but have adequate levels of methionine and cystine. Beans are high in lysine but low in methionine and cystine. Put them together in the same meal—supplementary proteins must be eaten at the same time—and you have a complete protein that is as good as meat or dairy products. The traditional Middle Eastern combination of chick peas and wheat and the Pakistani-Indian mixture of lentils or other beans with rice, wheat, or some other grain offers a similar balance. According to Lawrence Kaplan, the expert on ancient beans, the Iroquois Indians had their own version of supplementation: succotash. This mixture of corn and shelled immature beans offers a good balance of protein.

"I wonder how all these cultures figured out that certain combinations of vegetables were good for them?" I mused to an agricultural scientist recently.

"Maybe the ones that didn't figure it out died off," he answered.

The number of healthy vegetarians in various cultures is a good argument for the efficacy of supplementation but its value can be demonstrated more clearly in the laboratory. In a study carried out by the Guatemala-based INCAP (Institute of Nutrition of Central America and Panama), a combination of cowpeas and maize (corn) had a PER of 1.84 when each of the vegetables supplied half of the protein. Cowpeas alone had a PER of 1.41, maize alone 1.22. Other INCAP studies show that a combination of 90 percent rice and 10 percent black beans has a PER of 2.32 compared to 2.15 for rice alone; a 90–10 combination of maize and beans has a PER of 1.40 compared to 0.87 for maize alone; a 90–10 combination of sorghum and beans has a PER of 1.39 compared to 0.88 for sorghum alone; a 90–10 mixture of wheat and beans has a PER of 1.73 compared to 1.05 for wheat alone; and a 90–10 combination of oats and beans has a PER of 2.37 compared to 1.60 for oats alone. Milk casein, the major protein of milk and a frequently used reference protein, has a PER of 2.50. One way to bring a grain-bean diet up to the level of milk is simply to eat more beans. When INCAP researchers fed laboratory rats a diet of 50 percent black beans and 50 percent Opaque 2 maize, a special high-lysine variety (see Chapter 5), the PER of the combination was 2.60, higher than that of casein.

According to Ricardo Bressani of INCAP, studies show that the optimum combination of black beans and maize, the most popular grain in Latin America, is 78 grams (2.73 ounces) of maize and 28 grams (1 ounce) of beans—a 2.7 to 1 ratio. In actuality, however, adults in Latin America probably eat much more grain than beans. Surveys taken in Central America show that the ratio is about 12 or 14 to 1 in favor of maize. Figures like these indicate why the principle of supplementation, which keeps millions in good health, doesn't work for everyone. An added reason for the failure of supplementation is the quality of the food which the beans are used to supplement. In some developing countries, the staple vegetable—the one consumed in the largest amount— is cassava, known to us as tapioca. A delicious root, cassava contains, unfortunately, very little protein. A study shows that even when a basic diet of cassava flour is supplemented by 45 percent bean flour, the PER is only 0.96. Pushing bean flour up to 55 percent raises the PER to a still unsatisfactory 1.28. It isn't necessary, of course, to confine supplementation to plants with complementary amino acids. A little meat, fish, or dairy product added to legumes has the same effect in raising PER as a balancing of grains and legumes. Dr. Louis B. Rockland of the U.S. Department of Agriculture's Western Regional Research Laboratory in Berkeley, California, carried out a study which shows that a casserole of lima beans, meat, and milk has an impressive PER of 2.5 and is more digestible than meat alone. The reason for the increased digestibility of the legume-meat-milk combination isn't known.

Most of the nutritional shortcomings of beans can be overcome by supplementation and cooking, but another one, unfortunately, has defeated the best efforts of cooks and nutritional scientists. It's flatulence, a polysyllabic word for gas. An anonymous author once summed up the problem in this catchy jingle:

Beans, beans,
The musical fruit,
The more you eat,
The more you toot.

Flatulence has its funny aspects. There is a hilarious scene in Mel Brooks' movie *Blazing Saddles* in which a group of cowboys sits around the campfire emitting gas after a meal of baked beans. But flatulence becomes a serious problem when you consider the nutritional consequences in countries like our own. Dr. M. Wayne Adams considers flatulence to be the number one barrier to wider bean consumption on the part of the American public. Unless the flatulence-producing properties of beans can be reduced, he says, our bean consumption probably won't rise above the present low level of six pounds per annum per capita. At present, scientists really don't know how to get rid of those gas-producing properties or even what triggers them. The most widely held theory regarding legume flatulence is that certain bean sugars known as stachyose and raffinose are not digested in the higher portions of the human intestines, where most digestion takes place. The sugars descend to the lower portion of the intestines, where a microorganism, *Clostridium perfringens*, turns them into sucrose that can be utilized by the body. In doing so, however, it produces hydrogen, carbon dioxide, and, in some cases, methane, which are expelled from the body via the anus. It is these gases that comprise the phenomenon we call gas. Studies show that in humans, gas is not expelled until three to seven hours after the consumption of beans. Gas sometimes has an offensive odor, but not always, and some people, for unknown reasons, produce much more gas than others. In fact, some nations seem gassier than others.

"My Latin American students tell me they *never* have a problem with gas," says Adams.

The reason for the reported lack of flatulence on the part of certain populations may not be simply psychological, according to Adams. Perhaps, he suggests, the beans Latin Americans eat produce less flatulence. Or perhaps they have more or different microorganisms in their intestines that do a better job of digesting beans than ours do. There is evidence from other studies that different populations do have different bacteria in their digestive tracts. Or maybe Latin Americans just don't worry about gas very much. Gas isn't strictly an American phenomenon, though. One researcher who studied bean consumption by children in India, where beans are an accepted part of the diet, reported that the youngsters showed definite signs of flatulence after eating large amounts of beans.

For obvious reasons, measuring legume-induced flatulence production is one of the trickiest problems in legume research. A good laboratory assay eludes researchers, leaving them with the option of testing either animals or humans directly. Dr. Doris Calloway of the University of California, who has carried out a number of experiments on human flatulence, claims the easiest method of measuring gas in humans is to attach a collecting device such as a catheter to the anus via a tube, feed the subject beans, and sit back and wait for gas to be produced. The gas collected from the anus is analyzed by a standard laboratory instrument, the gas chromatograph.

By means of these and other techniques, researchers have learned a great deal about the gas-producing properties of beans, even if they haven't determined exactly what causes them. For one thing, they've found that some beans produce much more gas than others. Work by Louis B. Rockland indicates that blackeye peas, lima beans, and garbanzos produce the least gas among the ten different beans commonly eaten in the United States. Soybeans produced the most, although black beans and pink beans were also high. Green peas, Great Northern, small whites, and pintos were intermediate. But don't strike soybeans from your shopping list. Research conducted by Dr. Edwin L. Murphy, who, like Rockland, is with the USDA's Western Regional Research Laboratory, indicates that although soybeans produce somewhat more gas than green peas or limas, they produce somewhat less than dry peas and mung beans and a great deal less than pintos, kidneys, and small whites. The latter three were the gassiest beans in Murphy's study.

FLATULENCE FROM DIFFERENT LEGUMES

	RATIO *
Phaseolus vulgaris	
California small white	11.1
Pinto	10.6
Kidney	11.4
Phaseolus lunatus	
Lima, Ventura	4.6
Lima, Fordhook	1.3
Phaseolus mungo	
Mung	5.5
Glycine max	
Soya, Lee, or Yellow	3.8
Arachis hypogaea	
Peanut	1.2
Pisum sativum	
Pea, dry	5.3
Pea, green	2.6
Bland test meal	1.0

* Ratio of flatulence from test meal of 100 g (dry weight) for a three-hour period measured from four to seven hours after ingestion as compared to a bland test meal.

Source: Murphy, 1972.

A study by Doris Calloway, Doris Hickey, and Murphy bears out these findings in that it indicates that small whites are the gassiest beans among a group that includes soybeans, mung beans, lima beans, and the whites. The Calloway-Hickey-Murphy team came up with another interesting finding in their study. Soy and mung bean sprouts produced less gas than the whole bean, but still a substantial amount in comparison to the weight of the bean part (the cotyledon) consumed with the sprout. For some people, they point out, sprouts may present a digestive problem.

Murphy thinks some of the difference in the figures for whole beans reflects the fact that different varieties of the legumes in question were used. He found that one variety of lima bean, Fordhook, yielded one-tenth as much gas as another variety, Ventura.

Bean flatulence can be reduced, although all the methods of doing it have drawbacks. For instance, studies with dogs and people show that you can kill the organisms in your stomach that produce gas by taking antibiotics with your beans. But who wants to eat a dose of antibiotics with his bean soup? Processing is a more practical approach. Traditional Oriental processed bean foods (see Chapter 3) apparently produce little flatulence, indicating that a water extraction process such as that which produces tofu or a fermentation process such as that which results in tempeh, makes beans more digestible. Edwin Murphy of the WRRL has devised a modern counterpart of these ancient methods, an ethanol extraction process applicable to large-scale production. In one experiment, he turned white beans, perhaps the gassiest beans, into a virtually gasless product. Unfortunately, the flavor of the beans was lessened, a phenomenon that has occurred in other chemical extraction processes. Still another method of lessening flatulence is germination. Dr. Clifford Bedford of Michigan State germinated navy beans, killing off gas-producing sugars in the process. But the germinated beans have a small root growing out of them, raising the question of consumer acceptance.

Considering all the problems inherent in eliminating gas after it appears, some researchers think the best method of reducing flatulence is to deal with it *before* it gets started. Or, in other words, to breed a gas-free bean. There is a consensus among bean researchers that what is needed for the demanding American public is a better all-around bean with a superior balance of amino acids, no toxins, and, possibly, less flatulence. But breeding all these qualities into one super bean is a "formidable" task, according to Adams. It has already been learned that in beans, as in some other vegetable sources of protein, high protein is associated with low yields. Several studies, one of them by Adams, also indicate that high protein is, on the average, associated with low methionine and cystine. The cut-off point is apparently about 22 percent protein; beyond that, beans tend to be lower in the sulphur-containing amino acids, producing what Adams calls "empty protein." Adams suggests that the best strategy to achieve the super bean would be to narrow the search to high-yielding, medium-protein (20 to 22 percent) varieties. These, he believes, have the best chance of giving both high yield and high methionine and cystine.

Present yields in beans, he points out, are low by Green Revolution standards, the highest dry bean yield (excluding soybeans) being about seventy-five bushels per acre. In some areas of the world that desperately need protein, farmers are growing grain where they used to grow beans because they can get more money for bountiful grain crops.

"Yield is a nutritional problem," says Adams. "Beans have to be made to yield well or something else will be grown in their place."

Before more methionine and cystine can be bred into high-yielding beans, however, the trait has to be heritable. Some of the variation in protein levels is apparently due to environmental factors. Is there enough genetic variation in the sulphur-containing amino acids to breed higher levels into beans? The answer seems to be yes. In a 1967 study, two Michigan State researchers, Robert J. Evans and Selma L.

Bandemer, found that popular U.S. beans varied considerably in methionine and cystine. The highest level of both was in the black bean, which had a 3.0 score compared to a maximum 4.2 score for the FAO's ideal protein. More recently, Dr. John F. Kelly of Florida State University examined some 3,600 individual lines of *Phaseolus vulgaris* and found that they ranged from 0.8 to 3.9 percent methionine, a wide enough variation to indicate a genetic factor at work. The highest levels appeared principally in beans from China and the Caribbean. Ironically, one of the highest-ranking varieties turned out to be the mature dry Bush Blue Lake bean, a popular bean that is always eaten in the U.S. in an immature "green bean" stage. Later, Kelly found that the high methionine values in some of these lines were, indeed, inherited, although heritability is usually associated with low yield.

Kelly believes beans can be grown that will have twice the present content of methionine and possibly cystine, making unsupplemented beans almost as nutritious as eggs. Even an increase of about 25 percent in methionine, he says, could improve the general health in areas where beans are a prime source of protein.

A less flatulent bean is probably farther away. Dr. Carl Clayberg, a Kansas State University associate professor of horticulture and forestry, is in the early stages of research on breeding a gasless bean. When I visited him at his office at Kansas State University in Manhattan, Kansas, in 1976, he was literally surrounded by beans in bags, sacks, and bushel baskets. "Excuse the clutter," he said. "We're just finishing our bean harvest." The beans in the office, he said, represented crosses between pinto beans, the major bean grown in Kansas, and heirloom beans, a variety of *Phaseolus vulgaris* grown in the northeastern United States, and also known as Jacob's cattle bean. Among bean researchers, the latter is reputed to be relatively free of gas, although no actual tests with humans had been carried out at that point. Clayberg opened one of the bags, reached in, and took out some seeds with a purplish color. "This cross doesn't look like either of its parents," he said. Some of the others, he added, look just like pinto beans or heirloom beans, a white bean with purple markings. Why not just grow heirloom beans? Farmers, notes Clayberg, don't like them because yields and disease resistance are low.

He will plant the crosses, he explained, grow them, harvest the beans and, when he has enough, feed them to people to check their reactions. "Since gas comes out in discreet amounts, I think I can quantify it by the *number* of times it comes out," he says. His future plans also call for crossing the gassy pinto with the cranberry bean, another species of *Phaseolus vulgaris* reputed to produce little gas. Grown principally in Michigan, the cranberry bean is a pale tan bean with reddish-brown markings. Meanwhile, though, Clayberg is not waiting for a gasless bean. Since beginning bean research in Kansas (he is originally from the East Coast), he has developed a fondness for the state's major bean and has worked out his own version of pinto bean chili, which reverses the usual proportions of beans and meat.

The super bean is still in the future, but today in some areas you can buy an improved dry bean product that largely overcomes another bean drawback: cooking time. A quick-cooking bean, the new product should shortly be available throughout the United States. Cooking ordinary dry beans, as all cooks know, takes a long time. The quickest, lentils and split peas, take up to an hour; the slowest, soybeans, take up to *nine* hours. Most beans take two to three hours to cook, and that's after soaking. Incidentally, the United States Department of Agriculture recommends replacing the traditional overnight soak with a new method. Add six to eight cups of water to one pound of dry beans, bring to a boil for two minutes, then take the pot off the heat and let it stand, covered, for an hour. Even with this quicker soak, though, it takes so long to prepare beans that most vegetarian cookbooks, which stress beans heavily, recommend a pressure cooker to reduce cooking time drastically. A pressure cooker is a good idea if you're going to cook lots of beans. Pressure cookers are unavailable in developing countries, though, where the long cooking time of beans wastes precious fuel. Traditional cooking wastes fuel here, too, of course.

This particular problem, however, may already have been solved, at least in our own country. A decade ago, Louis Rockland of the WRRL in Berkeley worked out a method of pretreating dry beans which reduces cooking time drastically. The beans take just fifteen to thirty minutes to cook, depending on how they are processed, and require no soaking. The processing, which is carried out before the beans reach the consumer, involves three or four steps: (1) loosening seed coats by means of a vacuum process or by blanching in hot water or steam; (2) soaking in a solution of common salts; (3) rinsing; and (4) drying. The last step can be eliminated, resulting in a product that must be kept under refrigeration but which can be cooked in fifteen minutes. The redried beans require about a half hour to cook. The beans can also be frozen. Rockland figures that the fifteen-minute beans save 80 percent of the cooking time and 80 percent of the fuel used in cooking beans in the traditional way. The new product has exactly the same PER as traditionally cooked beans, but at least one of the products offers a bonus: less flatulence.

"We have evidence from human studies that large limas prepared the quick-cooking way have about one-half the flatulence of those prepared the normal way," Rockland told me.

The process not only works with limas, the first bean on which Rockland tried it and the one on which he has done the most research, but also with pinks, pintos, small whites, kidneys, blackeye peas, garbanzos, dried whole and split peas, and lentils. It even works on soybeans, although they take longer to cook—up to fifty-five minutes. Still, that's a big saving over the five to seven hours most cookbooks recommend for soybeans. Rockland also claims that his process improves the flavor of soybeans, diminishing the bitter, beany taste. For years after Rockland invented the process, it languished in the laboratory, but in 1976, a number of food firms suddenly put quick-cooking beans on the market. When I talked with Rockland in early 1977, refrigerated limas and blackeye peas prepared according to his process were being sold in the California area by four different firms. Two others were planning to introduce a variety of frozen quick-cooking beans, one of them to the consumer market. Eventually, the consumer-oriented firm planned to market the frozen product nationwide. A national chain also was considering offering the beans.

The price of quick-cooking beans is slightly higher than that of ordinary dried beans, but lower than that of canned or frozen beans, according to Rockland. He predicts a big future for his baby, now that it's finally on the market. "I think these beans will be marketed all over the United States in ten to twenty years," he says. "All the common varieties of beans will be processed this way."

It may be the first step in the comeback of the eight thousand-year-old domesticated bean, our best source of vegetable protein.

5

A PRACTICALLY PERFECT CEREAL

Oh beautiful for spacious skies,
For amber waves of grain . . .

—"America the Beautiful"

"THIS IS LANCOTA," SAID DR. JOHN W. SCHMIDT OF THE
University of Nebraska. He sprinkled a few dark brown seeds
from a jar into the palm of his hand. To me, the seeds looked
unremarkable in every way but I knew that they held a big
nutritional promise. Lancota is the world's first "super
wheat." A hybrid whose parents were a high-protein wheat
and two high-yielding wheats, it is both high in protein and
high-yielding, two traits that are difficult to combine. When
Lancota was grown experimentally throughout the world, its
seeds had from 1.1 to 2.3 percent more protein than the
average winter wheat and yields as good as popular wheats
grown in this country. The average winter wheat in this country produces about 13.8 percent protein, Lancota 15.3 percent protein. And all of its protein is located in the endosperm, the part of the grain milled for white flour.

"It makes a good loaf of bread, too," said Dr. Virgil A.
Johnson, the leader of the University of Nebraska research
team that produced Lancota. "Why, if we wanted just a high-

protein wheat, we would have had one in 1962. But we couldn't make it into bread."

What is needed for an acceptable high-protein wheat, he explained, is not only high protein and high yields, but good milling qualities—the qualities that turn wheat into a tasty loaf of bread or a high-rising cake. A higher than average lysine content is also desirable since wheat, like almost all the cereal grains, is low in this essential amino acid. With Lancota, the University of Nebraska researchers have all those qualities except high lysine but Lancota, with 3 percent lysine, has a shade more than the average Nebraska wheat. High-protein content in grain usually depresses the percentage of lysine. Lancota was officially released to U.S. farmers in 1975, and Johnson estimates that its production on all of

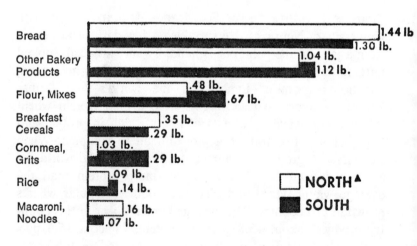

GRAIN PRODUCTS USED PER PERSON IN U.S.
Per Week by Region

Bread — 1.44 lb / 1.30 lb.
Other Bakery Products — 1.04 lb. / 1.12 lb.
Flour, Mixes — .48 lb. / .67 lb.
Breakfast Cereals — .35 lb. / .29 lb.
Cornmeal, Grits — .03 lb. / .29 lb.
Rice — .09 lb. / .14 lb.
Macaroni, Noodles — .16 lb. / .07 lb.

☐ NORTH▲
■ SOUTH

Quantities as purchased   ▲ Northeast, North Central, West
Households with incomes of $5,000-9,999 1 week in spring, 1965

Source: U.S. Department of Agriculture.

Nebraska's wheat acreage "could contribute as much as nine hundred railroad boxcars of additional protein in a single year."

We talked in Johnson's office, which has a sign on the door reading WHEAT IMPROVEMENT. It was August and 100 degrees outside on the university's Lincoln campus, but inside the buildings it was comfortably air-conditioned. Johnson, a tall, robust man with an open manner, was eager to talk about wheat, the focus of his research career. Lancota, he said, is a winter wheat (*Triticum aestivum L.*), as are all of the wheats grown in the central states and three-fourths of those grown in the United States. "They call it a winter wheat because it actually grows in the winter. You plant it in September, it grows all winter and it's harvested in July. Winter wheat *needs* cold weather. Good snow cover is important—it acts like a blanket." Throughout the world, he added, more winter wheat is growing than spring wheat, which is planted in spring and harvested in the fall of the same year. Winter wheats are used principally for making bread. Another kind of wheat, durum, is the basis for pasta products, such as spaghetti.

A young man put his head in the door to exchange a few sentences with Johnson. "He's from Pakistan," noted Johnson when the man withdrew. "We also have a student from Indonesia and one from Afghanistan doing their Ph.D. work here." Johnson, who has a research appointment with the U.S. Department of Agriculture's Agricultural Research Service, teaches no classes, but he supervises the work of doctoral candidates, some of them from developing countries where wheat is an important crop.

He and his colleagues in the Wheat Improvement group, Schmidt and Paul Mattern, believe the best way to deal with the protein shortage is to improve the world's most widely used form of protein, the cereal grains. Wheat and rice, Johnson points out in one of his research papers, are calorie and protein mainstays for two-thirds of the world's population. Corn and sorghum play the same role for a smaller number of people. Although cereal grains are usually eaten with another source of protein, such as beans, grain is the sole dietary form of protein for some of the world's people throughout part of the year. "In many areas of the world, people must live for months on wheat alone," said Johnson. Why not, the argument runs, raise the protein and lysine content of wheat to the point where it can supply all the protein the human body needs? This approach has the added advantage of dispensing with the expense and effort of introducing people to a new food. The food—wheat, in this case—remains the same; only its nutritional content is different.

The University of Nebraska research teams think a high-protein wheat will be most useful in developing countries, which depend heavily on wheat as a protein source, but they don't rule out a more important role for high-protein wheat here. "I think by the year 2000, we will be depending more on vegetable protein in this country," said Johnson. "We'll have to go more to cereal, because cereals are more effective producers of protein. If you take an acre of land, you can produce more protein on it by growing wheat than you can by growing forage to feed livestock." Wheat, Johnson notes in another research paper, has the lowest yield of the "big four" grains—wheat, rice, corn, sorghum—but the highest protein content, an average of about 12 percent.

The development of Lancota indicates that a wheat which supplies all the body's needs may someday be possible, but there are plenty of pitfalls along the way. Johnson has been working on high-protein wheat since 1955. "I settled on wheat because of all the traits in wheat that were important, protein was the least researched." Mattern had joined him in 1953, Schmidt in 1954. Johnson and Schmidt, both agronomists, handle the breeding and growing end of the research while Mattern, a biochemist, is responsible for nutritional and milling analyses. The original goal of the project, which was supported at first only by the state of Nebraska, was to raise the protein content of Nebraska wheat by breeding high-protein wheat varieties and stressing the importance of nitrogen fertilizer, which has been shown to raise protein content. At the time, some Nebraska wheat had sunk to 10 percent protein levels because farmers failed to apply nitrogen. Wheat below 12 percent does not have acceptable milling properties.

The Nebraska team got its biggest spur in 1963, when Dr. Edwin Mertz of Purdue University in Indiana and other Purdue researchers discovered high-lysine corn. The feat indicated that the protein content of grains could be manipulated. By 1966, the Nebraska University wheat project had support from the U.S. Department of State's Agency for International Development (AID) and the research team began looking not just for a high-protein wheat that would do well in Nebraska, but for one suitable for developing countries as well. As part of this expanded effort, Paul Mattern began analyzing seeds from the world seed bank maintained by the U.S. Department of Agriculture at Beltsville, Maryland, and Fort Collins, Colorado. It supplies seeds to agricultural researchers from a collection of some one hundred thousand different kinds of seeds from all over the world. Requests are filled only when a seed is unavailable elsewhere.

By 1976, ten years after the Nebraska wheat project had received the first seeds from the bank, Mattern had analyzed the protein and amino acid content of more than twenty thousand different wheat seeds. The protein content, he discovered, ranged from a low of 7 percent to a high of 22 percent—as high a percentage as hamburger!

You won't be able to leave all the hamburger out of your bun, however. Only about five percentage points of this wide variation in wheat protein is due to genetics. Still, that can supply a significant boost. In breeding tests conducted by the Nebraska team, a 12 percent protein wheat has already been raised four percentage points to 16 percent, a 25 percent increase. Lysine, unfortunately, offers less genetic potential for manipulation. Lysine content in the world wheat seed collection ranged from 2.2 to 4.0 percent, but only about 0.5 percent is attributable to genetics. It would take the maximum genetic boost to lift the few super-high lysine wheats to the 4.2 percent lysine standard set by the Food and Agricultural Organization of the United Nations for its "reference protein." The lysine content of most wheats is below 3 percent and could not be raised to 4.2 percent simply by genetic manipulation.

Lysine augmentation in cereal grains faces another problem: Increasing protein content usually lowers lysine content. It works this way in wheat, but the relationship, the University of Nebraska research team has found, doesn't lead to a less nutritious grain. True, the lysine content of the protein is lower in high-protein seeds, but the lysine content is actually *higher* when expressed as a proportion of the weight of the entire grain. Also, the relationship between lysine and protein that results in less lysine vanishes when the protein content exceeds 15 percent. No further decrease in lysine occurs beyond this point.

Feeding tests indicate that high-protein wheat is, indeed, more nutritious than standard wheat. When laboratory mice at the University of Nebraska ate diets containing 56 percent wheat, they gained more weight on two high-protein varieties than they did on standard wheats.

Genetic manipulation isn't the only factor that affects the protein content of wheat. Nitrogen fertilizer makes a significant difference. "It's no more possible to fix protein in wheat at some predetermined high level by breeding than it is to fix yields by breeding," Johnson told me. "The environment, particularly the fertility of the soil, has a big influence on protein content." The best way to raise the fertility in most wheat-growing areas is with nitrogen fertilizer. The same is true of corn-growing areas. The Green Revolution—the dramatic increase in corn and wheat yields that occurred during the 1960's—was due not only to new high-yielding varieties but to more water and fertilizer, particularly nitrogen fertilizer. Today, however, nitrogen fertilizer is scarce and expensive.

Where does the nitrogen shortage leave high-protein wheat?

It's a question being asked frequently today and Johnson has a ready answer for it: Still out in front of its competitors in the race to fill the protein gap. "At whatever amount of fertilizer you use, the high-protein varieties give more protein," he argues. In one growing test he described to me, an experimental high-protein wheat kept its two-percentage-point protein advantage over a standard wheat widely grown in Nebraska no matter what amount of nitrogen was used and, in fact, even when *no* nitrogen was used. Some other growing tests indicate that larger amounts of nitrogen increase the protein content of high-protein varieties even more than those of standard varieties.

The variation in protein and lysine content in wheats in the world seed collection indicates that Lancota is "just a beginning," in Johnson's words. The University of Nebraska team pins much of their hopes for the near future on the offspring of a cross between Nap Hal, an old high-protein variety from India, and a high-lysine experimental wheat from Washington State University. The cross has already produced progeny that combine high lysine (3.4 to 3.6 percent) with high protein (15.3 to 19.4 percent). Yields? Lower than average for popular United States varieties, but the Nebraska researchers hope to breed in high yield, too.

It's always possible, too, that a seed with a protein potential that exceeds that of the known high-protein seeds will turn up in some obscure corner of the world. The University of Nebraska's wheat laboratory regularly tests seeds supplied by agronomists from all over the world. Paul Mattern conducted me on a quick tour of the facilities, a crowded suite of rooms (the group was scheduled to move to larger quarters shortly) jammed with conventional laboratory equipment and items peculiar to wheat research. When new seeds arrive, Mattern explained, they are ground to a powder (two part-time assistants do nothing but grind samples), cooked, and hydrolized, procedures that reduce the seeds to their amino acids and other components. In this form, the samples are either analyzed by the amino acid analyzer or put through a number of chemical tests performed manually by technicians. The day I was in the laboratory in the summer of 1976, one of the amino acid analyzers was dutifully scanning the 19,885th sample from the 1975 samples the laboratory had received. "It's from Chile," said Mattern after consulting a log.

If the Chilean sample looks like a good prospect, it will be grown either in Nebraska or in a winter nursery at Yuma, Arizona. The second-generation seeds are collected and analyzed and, if the variety still looks promising, crossed with desirable partners.

When enough seeds from such a variety are available, they are brought to the laboratory, milled in a miniature mill about the size of a baby grand piano and baked into loaves and cakes. Like the mill, the food items are doll-sized. "This is our bread pan," said Mattern, smiling as he showed me a tiny pan about four inches long by two inches wide. He opened a freezer, inside which were a number of little loaves. "Nap Hal," he said, picking up a rather pathetic-looking loaf, quite flat, with a coarse texture. "It makes a miserable loaf of bread," said Johnson, who had joined us for the latter part of the tour. Even before Mattern baked the Nap Hal loaf, he knew it would fall flat. A machine called the mixograph, which stands in a corner of the laboratory, mixes flour samples and presents the operator with a readout called the mixogram. The mixogram for Nap Hal looked to me like part of an old Palmer penmanship exercise but to Mattern it indicated that the old Indian favorite would never make it in American-style bread.

Many of the seeds coming into the laboratory today are from an operation the Nebraska group organized called the Winter Wheat World Performance Nursery. It includes forty countries (every wheat-growing country except China). Researchers in participating countries grow wheats supplied to them by the University of Nebraska team, including some with high-protein or high-lysine potential. In 1972 and 1973, for instance, two of the thirty wheats grown were Lancota and Atlas 66, one of Lancota's parents. Atlas 66 had the highest level of protein of any of the thirty wheats, a mean of 16.7, but Lancota produced a very encouraging mean of 15.5 percent. Each researcher is supplied with a data book, which he returns along with samples. After the seeds are analyzed in Mattern's laboratory, the results are published as a research bulletin.

Nebraska's wheat research group also has organized three international wheat conferences in the past five years, the last one in Zagreb, Yugoslavia. The proceedings, which were edited under Johnson's direction, are so hefty I had them sent home rather than lug them on the plane. Because of conferences, the world nursery, and other research activities, Johnson and his colleagues have visited most of the wheat-growing countries in the world. In 1976, for instance, Johnson led a delegation of United States agronomists on a four-week tour of China's wheat-producing regions. A constant stream of visitors from those same countries troops through the offices of Wheat Improvement, seeking expertise on high-protein wheat. It all adds up to a lot of work and even the ebullient Johnson is a little tired.

"It must be exciting to visit so many foreign countries," I said wistfully as Johnson drove me to the airport for my short hop to another Midwestern state. "I'd like to stay home for a while," he said with equal wistfulness.

But noted agricultural researchers get little rest today. Dr. Edwin Mertz of Purdue University, whose discovery of high-lysine corn provided much of the impetus for the current work on nutritional improvement of grain, had just retired from Purdue when I visited there in the summer of 1976, but he was already slated to teach a course on proteins at another well-known Indiana school, Notre Dame. In August, he was still working in his modest Purdue office. Unlike most agricultural researchers, who tend to be large, suntanned men who look as if they could harvest a crop single-handedly, Mertz, a biochemist, is small and quick-moving. "How did I get into corn research? Well, I had been working on proteins and I understood corn was the most important crop in Indiana so I said, 'I'll work on corn.' " That was in 1946. Eighteen years later, he published with Lynn S. Bates and Oliver E.

Nelson a famous paper in *Science* describing the discovery of high-lysine corn.

"I just kept piddling along," Mertz described his first eighteen years in corn research. "If I had been in industry, they would have closed my project fifteen years earlier for lack of progress."

The breakthrough, he remembered, came in 1963 when Nelson sent Mertz the seeds of a number of corn mutants with a soft, floury endosperm. Placed on a light box, an apparatus with a translucent top lighted from below, the floury endosperm seeds were opaque, unlike the hard, translucent endosperm of standard corn varieties (*Zea mays*). Nelson, according to Mertz, had speculated that floury endosperms had more lysine than hard endosperms, which have about 2.5 to 2.8 percent lysine. The second floury endosperm mutant Mertz analyzed turned out to have a lysine content of 3.4 percent, although its total protein content was about the same as that of standard corn, a little less than 10 percent. Mertz called the seed Opaque 2. The secret of its high lysine content lay in the fact that less of its protein was made up of a fraction (a portion that can be isolated) called prolamine, which has few amino acids, and more was made up of those fractions which have a high percentage of lysine. Mertz' later work showed that the high lysine content of Opaque 2 is contained in the same gene that produced the floury endosperm and that it is heritable.

Most of the subsequent growing research on Opaque 2 has been carried out by the Mexican-based research organization called CIMMYT, which is largely funded by the Rockefeller Foundation. The results of their work are highly encouraging. The gene of high lysine is readily inherited, the plants respond well under a wide variety of growing conditions, and animal and human feeding trials indicate Opaque 2 is much more nutritious than standard corn. CIMMYT's biggest feat to date has been changing Opaque 2's floury endosperm, which is unappealing to people used to the chewiness of standard corn, to one more like the hard, translucent endosperm of familiar varieties without losing the high-lysine advantage. There are big hopes for the new variety, which was released to farmers in the United States, Brazil, and Mexico, all major corn-raising nations, in 1972.

As yet, though, Mertz admits, Opaque 2 has hardly revolutionized corn-eating habits. In Central and South America, where corn is a basic source of calories and protein, most people are still eating their old varieties of corn, which have an average of only about 2 percent lysine. The story is the same in the United States, which feeds most of its corn—80 to 85 percent—to animals. Figures show that Opaque 2 accounted for less than one-half of one percent of the total United States crop in 1972 and only about 4 percent of the total Brazilian crop. The biggest problem is yield. After a decade of work, CIMMYT has gotten Opaque 2's yield up to within 10 percent of that of standard corn, but corn is the highest-yielding grain and farmers are used to reaping bumper crops.

"Unless you pay farmers a premium for growing high-protein crops, they'll go for yields every time," says one agronomist.

Edwin Mertz, who spent eighteen years looking for high-lysine corn, is still optimistic about Opaque 2. High-lysine corn, he believes, will increase its yield and eventually win acceptance in areas where corn is a staple food. He also sees the United States turning more to high-lysine varieties, not only for animals but for people. "We'll all be eating more grains someday, especially the affluent areas," he predicted in 1976. At Notre Dame, he will work on the development of corn products for the American public. "We don't eat much corn in this country, but there are a lot of delicious corn products. *Arepas,* for instance, a little corn cake eaten in Colombia, are excellent. A frozen, high-lysine *arepas* made from Opaque 2 might be a very attractive food item."

With Mertz' departure from Purdue, no further research on Opaque 2 is in progress, but the Indiana institution is involved in the nutritional improvement of another important food

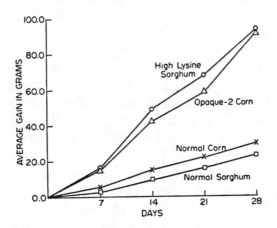

Weight Increase in Grams in Rats When High-Lysine Crops Are Used.

Source: *War on Hunger*, AID, 1976.

grain, sorghum (*Sorghum bicolor*), which is eaten by some three hundred million people in Africa, Asia, and South America. In this country, grain sorghum is an animal feed. A drought-tolerant species, it is grown principally on the hot, dry plains from Texas to South Dakota. In 1976, Dr. John Axtell, a geneticist who heads Purdue's sorghum research program, received the Alexander von Humboldt Award for the discovery of a high-lysine sorghum. It has approximately double the lysine of standard varieties when considered as a percentage of the whole grain. High-lysine sorghum and corn have a number of similarities. In both, the high lysine is associated with a reduction in the protein fractions poor in lysine and an increase in fractions rich in lysine. Another similarity is that a single easily inherited gene controls high lysine in corn and sorghum.

Mertz' work made the search for high-lysine sorghum easier

in that it indicated that floury-endosperm varieties were a good place to look, but Axtell's task was still herculean. Starting in 1966, when he first received a grant from the Agency for International Development (AID), he and his colleague examined almost ten thousand varieties of sorghum. Axtell's breakthrough, like Mertz', happened suddenly. In 1972, technicians analyzing two floury-endosperm varieties of sorghum collected in Wollo Province, Ethiopia, in the 1960's, found they had from 3.13 to 3.33 percent lysine. Standard varieties have an average of 2.05 percent lysine, the lowest for the big four grains. The protein content of the Ethiopian varieties was about average for sorghum—12.6 percent. The discovery that Ethiopia was the home of high-lysine sorghum seemed particularly fitting because it is the ancestral home of sorghum.

Were the two Ethiopian varieties an answer to increasing the lysine content of sorghum, just as Opaque 2 was the answer to increasing the lysine content of corn?

Apparently they were, but in 1973 there was no way to tell. The Ethiopian seeds had come from the world collection in India and only a few were available. To grow the seeds, Axtell needed more. So, in late 1973, he and Dr. Dallas Oswalt of Purdue flew to Ethiopia and, together with two Ethiopian agronomists, began the search for the high-lysine sorghum varieties. The notes of the original collector indicated that one of the strains came from just north of Dessie, the capital of Wollo Province. Driving along the dusty Ethiopian roads, Axtell and his party kept poking their heads out to check the appearance of sorghum in the fields beside the road. They stopped frequently to ask people about sorghum. "We found that we didn't have any difficulty talking about sorghum with the people," Axtell told me. "The people in that part of the country really know their sorghum. It seemed that everybody we talked to, including the children, could go through a field of sorghum and identify each different type. It's almost like American kids being able to identify cars by their make and year."

They found the first variety, *Mar-chuke*, which means

"honey squirts out of it," growing right beside the road, along with other varieties. It is highly regarded for its taste but yields are low, so it is used mainly to flavor dough prepared from high-yielding varieties. The second variety, *Wetet Begunche*, which means "milk in my mouth," took a little more searching but the Axtell party finally found it in a low-land village, Degan. The same area turned up still another floury-endosperm sorghum, which looks much like *Mar-chuke* and has since proved to have a high lysine content. Both Degan varieties are eaten together with other sorghum as a sort of flavoring. The taste, according to Axtell, resembles roasted chestnuts.

While Axtell was tracking down the source of the Ethiopian seeds, one of his graduate students, D. M. Mohan, was pursuing another route to more nutritious sorghum: chemical mutation. A process developed in the 1950's, chemical mutation involves producing a mutation—a sudden and profound change in structure that can be transferred by heredity—by means of chemicals instead of by waiting around for Nature to do the job. The chemical used in this case was diethyl sulfate, or mustard gas, minute amounts of which can induce mutations in seeds in just a few hours. After growing the treated seeds, Mohan screened some 23,000 sorghum "heads" —the plant's characteristic clump of seeds—by placing seeds from each over a light box and looking for opaque kernels. He found 475 likely-looking heads, 33 of which had at least 50 percent more lysine than standard varieties. The seed of one head, P-721, is extremely promising. Not only is its protein content higher than that of standard varieties—about 13.9 percent compared to 12.6 percent—but its lysine is a healthy 3.09 percent.

Rat-feeding trials with descendants of both the chemical mutant and the seeds Axtell brought back from Ethiopia indicate that both have about the same protein efficiency ratio, which is double that of the best United States standard varieties and triple that of the average variety.

In the course of his research on lysine, Axtell also unearthed another important aspect of sorghum nutrition. Certain lines, he discovered, are high in tannin, a substance found in a number of popular foods, including tea, grapes, and red wine. The high-lysine varieties, however, are low in tannin. In feeding trials, rats that ate high-tannin sorghum did poorly even when their diet was supplemented with extra lysine to bring the amino acid balance up to optimum standards. Rats that ate low-tannin sorghum supplemented with lysine, however, did almost as well as rats that ate casein, a milk product. Something in tannin, suggests Axtell, prevents rats—and, by analogy, people—from using all the protein found in sorghum. More than half the protein in high-tannin varieties may be affected, reducing the already meager protein intake of populations depending on them.

Exactly how tannin affects protein is not known, but Axtell thinks it may bind to some of the protein, making it unavailable. About 60 percent of all the sorghum varieties he has analyzed in the world collection are high in tannin, but in some areas, such as the Cameroon in Africa, almost all the sorghum varieties are high in tannin. High-tannin sorghums play an important role in certain areas because the sharp, astringent taste associated with tannin apparently discourages birds, a major menace to crops in developing countries. "In Ethiopia, kids sit on stands in the field and chase birds away," said Axtell. "Birds will completely devour a sorghum field in a half day if you let them."

The benefits of a high-lysine, low-tannin sorghum may play a significant role in alleviating world malnutrition. In the United States, grain sorghum is fed to animals, but sorghum is the staple food grain for millions in Africa and Asia and, to a lesser degree, in South America. Its popularity is due to its hardiness. Resistant to drought, it yields well and grows fast under the poorest conditions. "Sorghum will grow where no other grain will grow," said Axtell. He indicated on a map in his office the sorghum belt in Africa, which runs ten degrees north of the equator to ten degrees south. African sorghum products include unleavened bread, porridge, and a strong, dark beer. "In Ethiopia, they say a woman must be able to brew sorghum beer to make a good marriage," said Axtell, who has drunk the homemade brew without much enthusiasm. ("It's full of sticks.") Sorghum is also popular in parts of India and in some South American countries, such as Venezuela.

The major impact of high-lysine, low-tannin sorghum will probably be in countries where it is used as a staple food, but a more nutritious variety also would improve the sorghum feed used for United States livestock. Currently, sorghum feed is supplemented with high-lysine soybeans. A feed that supplied most of the nutrients needed by chickens, pigs, and other animals would free more soybeans for people. At the ceremonies at which Axtell received the von Humboldt award, Fred L. Atterson, president of the American Society of Agronomy, pointed out another possible use for high-lysine, low-tannin varieties of sorghum grown in the United States: an export crop for food use that would compete with less nutritious sorghums from other areas of the world. At present, most of the sorghum we export is fed to animals.

Before any of this happens, though, Axtell and his group have a number of difficult research problems to work out. A high-lysine, low-tannin sorghum is a reality but a high-lysine, low-tannin variety that yields well and has a hard, translucent endosperm like that of standard varieties is still in the future. Cross-breeding of strains with the desired qualities will, it is expected, eventually produce such a line. A more problematical goal is increasing the lysine content of sorghum endosperm still further by breeding out much of the prolamine, the protein fractions that contain the fewest amino acids. Decreasing prolamine from the present 50 percent of the endosperm to about 10 percent, Axtell figures, could bring the protein quality of sorghum, now the least nutritious of the four leading cereals, up to that of the most nutritious, rice. Increasing the size of sorghum's unusually small germ, the most nutritious part of the seed, also may lead to increased lysine.

With all these goals to reach, Axtell's laboratory is full of activity, even in late summer. Technicians analyze sorghum seeds for promising mutants in one room. During my visit, a young woman seated at a table was using a small, boxlike machine called a grain quality analyzer to determine the protein level of a seed grown in Purdue's Puerto Rico winter nursery. "It's 11.8 percent," she said in response to Axtell's question. A larger room functions as a sorghum bank. "There are about sixteen thousand varieties of sorghum in the world and we have samples of twelve thousand of them here," said Axtell. He pulled out a card at random from a file box. The card contained a transparent packet with seed, a number, and a notation on the area where it was collected. A light box is set against one wall.

"Do you want to see how it works?" asked one of Axtell's graduate students, a Lebanese named Vartan Guiragossian. He sprinkled some of the high-lysine Ethiopian seed on the lighted top; no light passed through the opaque endosperm. He removed them and put down some standard seed; the light shone through.

Before I left Purdue, Axtell drove me out to the university's research farm, which includes twenty-one acres of sorghum varieties from all over the world. Axtell is well over six feet tall, but the high-protein Ethiopian varieties he has growing in one area topped him by several feet. "They're about eight feet tall now but they'll keep on growing to twelve feet— right to the first frost," he said. "In Ethiopia they use the stalks as a building material." At the top of each plant were bushy heads of small seeds. Each variety had a different color seed: red, brown, white, cream, orange. In other areas of the sorghum section, the plants were considerably shorter than the Ethiopian varieties. Sorghum, explained Axtell, was brought from its original home in Africa to South America by African slaves and from there to our southwestern states. Gradual selection of shorter varieties over the years turned it into a plant about half the size of its towering ancestor.

"Did you ever have sorghum syrup on pancakes?" he asked.

I admitted I had never even heard of this treat, which is

made, Axtell explained, from a special variety of sweet sorghum. "It's popular in the South and Midwest," he said. He promised to stop at a local supermarket so I could buy a jar. As another memento, he and Kay Porter, the assistant director of the sorghum project, picked a small bunch of some of the more colorful sorghum heads and presented it to me. "My wife and some of the other people around here use these as a decoration in the fall, when the color turns deeper," said Axtell. I bought a jar of sorghum syrup and carried it and the handsome bouquet of sorghum home on the plane. The bouquet caused a mild stir.

"Uh—corn?" asked one man.

"Close—it's sorghum."

"Beautiful!" said one woman. "Where can I get some?"

Edwin Mertz' work on high-lysine corn has inspired still other successful efforts at genetic improvement of a food grain. In 1970, a group of Swedish researchers published a paper describing the discovery of the first high-lysine barley, "Hiproly," which, like John Axtell's sorghums, originated in Ethiopia. In the case of barley (*Hordeum vulgaae*), the discovery process was a little easier; the investigators analyzed only two thousand samples from the world collection before finding Hiproly. It has about 30 percent more lysine than normal barley. Shortly after the discovery of Hiproly, some promising artificially induced mutants were produced by workers in neighboring Denmark, one with 45 percent more lysine than standard varieties of barley. The artificial mutant and Hiproly both have a better biological value (BV) than standard barleys.

Oats, another minor grain, also may be amenable to genetic manipulation of their protein content. Oats start off with one big advantage. On the basis of protein content and amino acid balance, oats are one of the world's most nutritious grains. *Avena sativa*, the species grown commercially in the United States, has an average of 17 percent protein. Many varieties of the species have more, as do some other species of oats. In a survey of eleven species of *Avena sativa* grown throughout the world, Yeshajahu Pomeranz, Vernon L. Youngs, and G. S. Robbins, all with the United States Department of Agriculture's Agricultural Research Service office at the University of Wisconsin (Pomeranz is now with the ARS' Grain Marketing Research Center, Manhattan, Kansas), found that the maximum protein content of the groats (dehulled oat seeds)

PROTEIN CONTENT AND ESSENTIAL AMINO ACID
COMPOSITION OF GROATS FROM 11 OATS SPECIES

PROTEIN OR AMINO ACID	MAXIMUM (%)	MINIMUM (%)	MEAN (%)
Protein	37.1	17.8	27.1
Lysine	4.1	3.5	3.8
Threonine	3.4	3.1	3.3
Cystine	2.5	1.9	2.3
Valine	5.8	5.2	5.5
Methionine	3.3	2.2	2.9
Isoleucine	4.1	3.8	3.9
Leucine	7.6	6.9	7.3
Tyrosine	3.5	2.7	3.2
Phenylalanine	5.5	5.0	5.3

Source: Pomeranz, Youngs and Robbins, 1973.

---

was an almost unbelievable 37.1 percent. The minimum was 17.8 percent. *Avena sterilis,* a wild species, had a range of 22.1 to 31.4 percent protein. The maximum lysine content of the eleven *Avena* species was 4.1 percent, the minimum 3.5 percent. The maximum is just below the 4.2 percent lysine standard set for the FAO reference protein. Earlier, Robbins and Pomeranz found a maximum of 5.2 percent lysine among almost three hundred *Avena sativa* seeds they examined. In oats, unlike wheat and sorghum, increased protein content doesn't mean lowered lysine, another nutritional advantage for this cereal grain.

Unfortunately, all of the eleven *Avena sativa* species and

the wild *Avena sterilis* have what the researchers call "poor agronomic characteristics," which means American farmers are not likely to plant them. But crosses with our commercial lines may result in oats that not only yield well but offer a protein content similar to beans and an amino acid balance almost as good as the FAO reference protein. A practically perfect cereal, in other words.

A program to produce better oats is already underway in this country. Oat seeds from promising lines of *Avena sativa* and *Avena sterilis*, as well as crosses between various lines, are distributed to breeders, and samples from each year's harvest returned to the Oat Quality Laboratory at the University of Wisconsin. In 1976, the laboratory analyzed 24,425 samples; in 1977, it expected to analyze 40,000. Two new high-protein oats are now available to growers in the United States, one of them, Dal, released by the University of Wisconsin. It has as much as 22.3 percent protein, dry weight, although it averages around 20 percent in production. The lysine content is about 4 percent. Large oats buyers are now paying a premium for high protein content, according to Vernon Youngs, ensuring the continued planting of the new varieties. Although most oats are used for livestock today, he thinks the high-protein varieties will lead to increased human consumption of the nutritious grain.

Does this mean oats will replace wheat? No. Oats have no gluten, the substance in wheat that helps bakery products rise, so it is impossible to make an acceptable loaf of bread out of oats alone. Another drawback to oats is their low yield, even in commercial varieties. But we can do a lot more with oats than put it in cookies and cereal, and it seems likely that oats will show up in more and more foods in the future. One possible use is in fortification of soft drinks, a project on which researchers are working at the ARS's Northern Regional Research Laboratory in Peoria.

We don't use much barley or oats as human foods in the United States, but rice, a popular grain here, also looks like a good prospect for genetic manipulation of its protein content. Domesticated brown rice has the highest lysine content of the major grains, 4 percent, but the lowest protein content, about 7.5 percent. White rice has only 6.7 percent protein. The International Rice Research Institute, which is head-quartered in the Philippines, has set a goal of a 2 percent increase in the protein content of rice. Already, the Institute has lines in which the protein content ranges from 1.7 to 4.4 percentage points above that of a popular high-yielding variety used as a standard. Yields and lysine content are lower, just as they were in the first high-protein wheats. But enough variation in those two factors has been found to make rice researchers confident that crosses offering high protein, normal lysine, and high yields can someday be produced.

Familiar plants aren't the only source of high-protein grain. After a history of almost one hundred years of research, a successful hybrid of wheat and rye was produced in the late 1960's by the University of Manitoba in Canada. It was called triticale, a combination of the scientific names for wheat (*Triticum aestivum*) and rye (*Secala cereale*). One variety of triticale, Number 204, has 17 percent protein. Two others have just under 14 percent. Unfortunately, this protein content decreases when the grain is milled for flour. Number 204 flour, for instance, has only 14.1 percent protein. Still, that's more than most wheat flours, excluding the high-protein varieties. Triticale yields well, although not as well as wheat, and the bread-baking qualities of Number 204 are fairly good. In a series of tests run by Dr. Klaus Lorenz of Colorado State University, it did as well in bread as wheat flour when certain modifications in the bread-making process were made. What does triticale bread taste like? A "mild rye bread," according to Lorenz, who says most people who have tried it like it better than wheat bread. We may get a chance to taste triticale bakery products before long. Commercial varieties of triticale are now available to United States farmers and it's estimated that several hundred thousand acres are being planted with triticale here.

The goal of the "perfect cereal" has never been so close.

6

IT AIN'T (JUST) HAY

"Hey!"
"Hay is for horses, not for people."

—CHILDREN'S SAYING

IN HIS WHITE LAB COAT, DR. MARK STAHMANN, A PROFESSOR of biochemistry at the University of Wisconsin, looks like a filmmaker's idea of a scientist. A tall man with a full head of gray hair, he has the intent gaze of the visionary. For almost twenty years, Stahmann has been engrossed in research on the extraction of protein from green leaves, a concept he believes can solve the world protein shortage. Talking with him in his small office in the Department of Biochemistry on the University's Madison campus, I caught something of his fervor. "Agriculture is very backward," he began in rapid, fluent speech. "If your great-grandfather came back to earth, the place he would feel most familiar is at the dinner table. We are eating basically the same foods we ate then. If we're going to feed the world, we will have to make breakthroughs similar to those in other fields. Green leaves are the major source of all protein and carbohydrates, but leaves have so much cellulose that only ruminant animals can digest them. To increase our food supply, we must learn to use all the plant, including the leaves."

The existence of leaf protein has been known for two centuries, Stahmann continued, but research on the subject really started during World War II, when British scientist N. W. Pirie suggested the use of leaves to augment dwindling meat supplies. "I have a copy of Pirie's original proposal," said Stahmann, opening a drawer and taking out a sheaf of Xeroxed pages. Pirie's proposal never got underway during the war because of the costs involved, but after the war he was given a laboratory where he carried out most of the pioneering work on leaf protein. At the time, amino acid analysis, the backbone of protein research, was still being carried out by tedious and time-consuming manual methods. "The instrumentation to analyze amino acids was worked out only in the 1950's," said Stahmann. "Our laboratory had the first commercial amino acid analyzer ever made."

"I'd like to see it."

Stahmann led the way into the big laboratory next to his office. Standing next to one of the laboratory tables was the analyzer, a refrigerator-sized machine with a face full of dials. The laboratory now has two additional machines. Each costs about $60,000 today.

"One of the first things I did when I got this analyzer was write to Pirie," said Stahmann, looking at the machine with obvious pride. "Later I visited him and got samples of leaf protein. I wrote the first paper that described the amino acids in leaf protein. At this point, I realized that leaf protein looked like a good protein, but that amino acid analysis isn't enough. You can analyze hair and get a pretty good balance. The important thing is, how do animals utilize amino acids?"

Biochemists have developed a number of methods to measure protein utilization (see Chapter 4) but Stahmann worked out his own, which involves the digestive enzymes. This method, as well as other measurements, indicates that a derivitive of leaf protein is a well-balanced protein, containing ample amounts of all the amino acids except methionine and, to a lesser degree, cystine. Its PER—the weight gain in rats per gram of protein ingested—is 1.7. More refined derivatives have a PER of 2.40. The PER of milk casein, the most-used reference protein, is 2.5.

PROTEIN CONTENT OF SOME
LEAFY GREEN VEGETABLES

VEGETABLE	PROTEIN CONTENT (%)
Chard (raw)	2.6
Chard (cooked)	1.4
Dandelion greens (cooked)	2.7
Kale (raw)	3.9
Mustard greens (cooked)	2.3
Parsley	3.7
Spinach (raw)	2.3
Spinach (cooked)	3.1
Turnip greens (raw and cooked)	2.9

Source: *Human Nutrition,* by Benjamin T. Burton. Copyright © 1976 H. J. Heinz Company. Used with permission of McGraw-Hill Book Company.

These figures, it should be noted, apply to *fresh* leaves, which contain 20 to 30 percent protein on a dry-matter basis. Leaves that dry naturally lose much of their protein. If you go out and pick a mess of fresh spinach or kale for dinner, though, you won't be eating 20 to 30 percent protein, because most of the content of fresh leaves—upward of 80 percent—is water. Most fresh leaves we eat contain only 2 to 3 percent protein. Raw kale has one of the highest protein contents, 3.9 percent. Popeye's favorite, raw spinach, has 2.3 percent.

Most leaves have a similar balance of amino acids but not all leafy crops yield the same amount of protein. For this country, Stahmann found early in his research, the best bet is alfalfa, a foot-high plant with round, dime-shaped leaves that is the source of about half the "hay" consumed by ruminants such as the cow and the horse. The smell of new-mown hay writers talk about is really the smell of fresh-cut alfalfa. In an analysis Stahmann made of crops in the United States, alfalfa gave three hundred pounds of essential amino acids per acre, one hundred pounds more than its closest competitor, soybeans. The whole corn plant—seeds, leaves, stalk, and all —was third, followed by the whole soybean plant, and clover. Crops like alfalfa and clover that are grown almost exclusively for animal feed are called forage crops. In general, forage crops have much higher yields than food crops and can be grown on much poorer land, making them extremely attractive sources of leaf protein.

Alfalfa, however, has a few added advantages. It is a legume that fixes its own nitrogen, thus saving the farmer the expense of nitrogen fertilizer; it is a perennial that requires seeding only every four to ten years; it grows fast and recovers quickly after harvesting; and it thrives in a variety of locations. Stahmann thinks alfalfa alone could supply the world with all the protein it needs. In one of his scientific papers, he writes that an area a little bigger than Texas can supply enough alfalfa leaf protein to give every human being presently living on earth 35.2 grams of protein per day, the average requirement recommended by the Food and Agricultural Organization of the United Nations. To the world's farmers, the advantages of alfalfa as a crop are not news. A native of Turkey and Afghanistan, alfalfa is the leading forage crop in the United States and in many other countries as well. According to the FAO, it grows on some 6.5 billion acres throughout the world. In the United States, an estimated twenty-seven million acres are planted in alfalfa.

Most alfalfa is eaten by ruminants either as silage (fermented, semidried plant matter) or hay (dried plant matter). As a result of the American farmer's dependence on alfalfa as a forage, a substantial processing industry already exists to handle it here, giving the crop another advantage over other leaves.

There's just one drawback to alfalfa as far as people are concerned: We can't eat it. Like most leaves, the alfalfa leaf has a high content of cellulose, the source of fiber. Some fiber in the diet is fine, but too much in a plant makes it indigestible. To digest high-fiber plants requires the efficient multi-stomach digestive system of the ruminant animals, which can process cellulose into protein that can be utilized by the animal's body. Man, a single-stomached creature, can digest only those parts of plants that contain comparatively little fiber: seeds, roots, fruit, and the tender young leaves of some plants. Leaf protein from the high-yielding forages is reserved for ruminants. Or it was until recently.

The significance of the work by Pirie and other British agricultural scientists in the 1950's is that they made all of the world's leaf protein available to man for the first time. Machinery they developed ruptures the tough-fibered plant cells of leafy forage plants, such as alfalfa, and presses out a green juice that has 25 to 35 percent protein. When the juice is heated, it separates into curds and a clear brown liquid. The dried curds become a dark green powder: leaf protein concentrate or LPC. It contains from 40 to 50 percent protein and has a PER of 1.7. Rat studies show that the biological value (BV), or digestibility, is surprisingly high for a plant protein source: about 80, compared to 83 for milk, 80 for meat, and an average of 75 for soybeans. The fibrous press cake left when the juice was extracted in the first step contains about 18 percent protein, making it suitable as an animal feed.

The few scientific studies that have been carried out with humans and green LPC bear out the high BV obtained with rats. In India, Dr. Narendra Singh added 15 grams (.53 ounce) of LPC to 350 grams (.77 pound) of Ragi flour, a staple in India. When baked goods made from the flour were fed to Indian children for six months, they gained twice as much weight and showed twice the height gain of children who ate the same diet without leaf protein. A Nigerian researcher, Dr. O. L. Oke, supplied mothers of twenty-six severely malnourished children with LPC, giving them instructions to feed each child one tablespoonful in his usual food. Within ten days, the swelling associated with the malnutrition had disappeared and the children became more alert. A biochemical analysis indicated that the protein level in their blood had increased. Oke estimated that it would cost about one cent a day for enough concentrate to supply a family of six.

At this writing, a large-scale trial with leaf protein involving some six hundred children is underway in India. The study, which will run for two years, will compare leaf protein with skim milk and a local variety of legume.

From all indications, leaves supply a high-quality protein, but the nutritious food has had a struggle gaining scientific acceptance. One reason is cost. When Pirie and other British researchers first worked out the process to retrieve protein from leaves, the cost of soybeans was only about two dollars per bushel, a price which held throughout the 1960's. Today soybeans cost over four times as much. As a result of low soybean prices in the 1950's and 1960's, soybean protein cost far less than leaf protein during this period. One of the few American researchers who saw promise in leaf protein at that time was Stahmann. He visited Pirie in 1961, returning to the University of Wisconsin with alfalfa protein samples and plans for a protein extraction unit. The machinery, built with help from the state dairy industry, was soon pressing out leaf protein. The choice of leaves was a natural: alfalfa. Wisconsin is the nation's leading producer of alfalfa, which is eaten by the state's dairy cows. One of the biggest breakthroughs to emerge from Stahmann's alfalfa research was the discovery of how to preserve the press cake in silos as silage, the farmer's cold weather standby for cows.

"We put it in plastic bags, which produces more rapid fermentation," he explained. "Cows eat it and thrive on it. It's easier to make than regular silage."

In spite of this and other advances, financial support, as Stahmann remembers now with some rancor, was minimal for the first ten years he worked on leaf protein. During most of that period, he had funds to pay for the help of only one part-time technician. Two parallel developments, the rising price of soybeans and the growing protein shortage, have changed that picture radically. Leaf protein has become an attractive alternative form of protein. In 1975, Stahmann's research began receiving funds from the prestigious National Science Foundation. Two years before that, the University of Wisconsin undertook a large multi-disciplinary research effort involving eleven departments and fifteen faculty scientists. Stahmann and his colleagues now envision leaf protein being produced by a farm process in which individual farmers will use machinery developed particularly for that purpose at the University of Wisconsin. The process is also adaptable to large-scale central production.

One of the products that emerges from the farm process is green LPC, which has a protein content of about 44 to 50 percent. A second is the press cake, which has a protein content of slightly over 17 percent. A third product, the brown liquid left over after the green liquid is congealed into high-protein curds, contains nitrogen and potassium and is suitable for fertilizer or possibly a culture for the production of single-celled protein, such as yeast. The Wisconsin researchers see the green LPC, which can be digested by single-stomached animals, as a protein supplement for such nonruminants as pigs and chickens. It could replace part or all of the present supplements, which are based primarily on soybeans. By using the system, Stahmann emphasizes, a farmer can produce much more protein—estimates run up to three times as much—than he can by simply cutting his 20 percent protein alfalfa and feeding it to his cows as hay or silage. Not only that, but one of the products of the farm process, green LPC, can be eaten by nonruminants, including one particular nonruminant—man.

The machinery to produce all these products won't be cheap but Dr. H. D. Bruhn, an agricultural engineer and one of the directors of the alfalfa protein project, figures that the total return will put the farmer a little ahead, even considering the investment in machinery. At present, it is estimated that a unit to extrude, press, and coagulate protein will cost about the same as any other large piece of farm equipment. None of the machinery is available today, and in 1977 the Wisconsin team was projecting a delay of five years in putting it on the farm. Several mechanical problems remain to be solved, including the development of more efficient replacements for the pressing and coagulating operations, both of which use excessive amounts of energy.

I was at the University of Wisconsin on a cold November day, too late in the year to see alfalfa being processed, but Stahmann arranged for me to see a film of the process in Bruhn's office. "Alfalfa's one of the main crops here in Wisconsin," said Bruhn, as we watched a large red harvesting machine cutting fresh alfalfa and blowing it into the machine. "We have three million acres." On the screen, the chopped alfalfa was placed in an extruder, a machine the Wisconsin team finds uses less energy than conventional rollers. The extruder squeezed the alfalfa out through small holes, breaking the cell walls in the process. A mushy green material flowed into another machine, the screw press, which squeezed out the juice. "We have a control panel for this whole process," commented Bruhn. "It's largely automatic." The juice could be seen flowing into tubes where it was steamed to make it coagulate. The screen showed the result: a lumpy mass, bright green in color, and a brown fluid. "Looks like green cottage cheese, doesn't it?" asked Bruhn. "Now the protein concentrate is being dried on big rollers."

He turned off the projector and picked up a large jar of the end product, a very dark green powder. It gave off a strong grassy smell when he opened the lid.

LPC, he explained, can be produced right on the farm or sent to a central processing plant in juice or coagulated form for further processing. "The real advantage of this process as far as the farmer is concerned is the silage," he noted. "You need one day of sun to dry alfalfa after you cut it to make silage, two days of sun after you cut it to make hay. But with this process, you can cut alfalfa whenever you want—even if it's raining. It's weather-independent. That's important because you can come up with the nicest protein in the world and it won't do you any good if the farmer doesn't want to grow it."

Feeding trials conducted at the University of Wisconsin in the last few years show that the feeds produced by the farm process compare well with standard feeds. Cows that ate the alfalfa press cake left over after the LPC is extracted gave only slightly less milk than cows that ate conventional alfalfa silage (20.6 kilograms compared to 21.3). Nonruminants did well, too. Pigs that had 5 and 10 percent LPC in their diets gained just as much weight as pigs that received all their protein from soybeans and corn. Young chickens that ate a ration including 10.6 percent LPC supplemented with lysine gained nearly as much weight as chickens that ate a standard diet of corn and soybeans. Results such as these, claim the University of Wisconsin researchers, show that LPC and alfalfa press cake can be used to provide part or even all of the protein farm beasts such as these need in their diet.

Pigs and cows may like the new alfalfa products, but how about human beings?

When Bruhn showed me the dark green LPC and I caught a whiff of its strong, grassy smell, I had doubts about its acceptance by humans, particularly in this country. If most Americans reject soybeans because of a grassy, bitter taste, how much more likely are they to reject a product that is not only much grassier, but green in color? Green LPC, it has already been demonstrated, imparts a moldy-looking greenish tinge to baked goods that effectively restricts its use to one occasion in this country: St. Patrick's Day. True, green LPC has been used successfully in human diets on an experimental basis, but in every case it was given to undernourished young children in developing countries, a group with little choice in foods. One solution that has been advocated is putting LPC in highly flavored dark-colored foods (alfalfa-flavored ginger-bread?), but most nutritionists pin their hopes for LPC in human diets on getting the grassy taste and color out of the powder.

A group of researchers at the USDA's Western Regional Research Laboratory in Berkeley is devoting much of its attention to taking the green out of leaf protein. They already

ESSENTIAL AMINO ACID COMPOSITION OF WHITE
ALFALFA LEAF PROTEIN CONCENTRATE

AMINO ACID	LEVEL
Isoleucine	5.3
Leucine	9.4
Lysine	6.4
Methionine + Cystine	4.3
Phenylalanine + Tyrosine	12.0
Threonine	5.7
Tryptophan	2.5
Valine	6.9
Crude protein (dry basis)	89%

Source: Bickoff *et al.*, 1975

have a highly promising 90 percent protein white derivative of green LPC that they call Welpro. It has a higher level of methionine than green LPC, 2.3 percent compared to 1.9 percent, which brings it up to the suggested methionine level in the widely used reference standards set up by the FAO. A chemical "score" in which the essential amino acids in a food are compared with the FAO standards for all the essential amino acids gives Welpro a score of 100 percent. Only hen's eggs and beef muscle do as well. Cow's milk scores 95, casein 91, and soybeans 74. Welpro's PER is impressive, too: 2.40. With methionine supplementation, it's 2.70.

The taste? Bland, claims the Berkeley group.

Like the University of Wisconsin farm process, the Berkeley process starts with fresh chopped alfalfa. The material is fed through a screw press, which ruptures cells and presses out juice, leaving behind the press cake. The juice is heated, just as in the Wisconsin process, but at a point below that at which the juice would coagulate, the heat is stopped and the juice cooled. Before long, the cooled juice separates into two phases, a green sediment and a clear brown liquid. When the brown liquid is heated, a flaky white solid appears. Spray-dry it and you have white LPC. Adding sodium metabisulfite to the product lightens it further and increases the levels of sulphur amino acids.

One of the big advantages of the new Berkeley process is that the animal feed products which give alfalfa protein its economic viability are still available. By heating and drying the green solids, you get the same 50 percent protein green LPC you would have had if you had continued on with the original Berkeley process. There's simply less green LPC, because part of it went into white LPC. The same amount of 18 percent protein press cake remains in both processes. The white LPC represents only about 3 percent of the original fresh alfalfa (dry weight), but it contains about 43 percent of the original soluble protein. The rest is in the remaining products. One of the feed products that emerges from the Berkeley process has already been produced commercially, although not primarily as a nutritional supplement. A California firm made green LPC under the name X-Pro and sold it as a protein concentrate to enhance the yellow color of chicken skin and egg yolks. Two alfalfa pigments, xanthophyll and carotene, produce the yellowing effects. The firm went out of business for economic reasons that have nothing to do with X-Pro, but a French firm is now making green LPC as an animal feed and two British firms plan to start making it.

Today green LPC for animals, tomorrow white LPC for people? Leaf protein might follow that scenario, just as soybeans have. The high-protein white powder is still so new that not much work has been done with utilization, but the Berkeley group suggests it will be used as a nutritional supplement in such foods as snacks, gravy, soup, pasta, milk substitutes, and meat extenders, in much the same way that soy protein is used today. The additional processing needed for white LPC makes it at least as expensive as soybean protein, but the Berkeley researchers expect the cost to drop in the future as a result of further research. "After all," one points out, "look how much money has been spent on soybean research over the past thirty years. Alfalfa protein research is just getting started."

Adding a new substance to conventional foods, even in small amounts, isn't a matter of simply throwing in a few tablespoonfuls. A successful additive must either have no deleterious effect on a food's functionality—the qualities that make it stick together, expand, remain in solution, or otherwise maintain its structural integrity—or actually aid it, as soy protein does in many processed foods. There is some evidence that alfalfa won't be as easy to use in food as the versatile soy. "In bakery goods, alfalfa is like sand," says Dr. Lowell Satterlee, a University of Nebraska researcher. "It's nutritional but not functional." Satterlee produces a light-colored alfalfa protein in the same way as the Berkeley group but his product has less protein and a slightly grassy flavor. The Nebraska researcher's findings do not relegate alfalfa to a back shelf in the food cupboard. He has come up with a concept that enables not only alfalfa, but other alternative proteins as well, to be used in some conventional recipes. Using a computer, he prepares blends of food protein for specific functional and nutritional properties.

Satterlee holds up a small bottle of yellowish powder which he calls APC—alfalfa protein concentrate. "Add this to cookies and it won't work. But a cookie with a mixture of alfalfa protein, whey, soy protein, and wheat protein concentrates has color, width, nutrition, and all the other qualities you want. And we can do this for bread, pasta, snacks, cereals, and sausage."

To get optimum blends, Satterlee codes data on amino acid profiles, costs, and other information into a computer. Then he programs it to select combinations that will result in a high-protein mixture with an essential amino acid profile comparable to that of the egg at the least possible cost. Thus far, the most successful item produced by the method is a cornmeal-based snack enriched with a high-protein flour containing 45.5 percent alfalfa. The snack, a puffy, extruded product that looks like the popular corn curl, was acceptable to a taste panel at all levels of fortification. At the highest level, it had a two to one ratio of cornmeal to high-protein flour, and a 23.2 percent protein level. The other ingredients in the flour are soybeans, dried potatoes, and cottonseed protein.

Alfalfa is getting most of the attention from leaf protein researchers in the United States today, but other leafy plants may also turn out to be good sources of protein. Mark Stahmann and two colleagues at the University of Wisconsin extracted green LPC concentrate and press cake from a number of plant wastes, including beet tops, lima bean vines, carrot tops, pea vines, and potato vines, all of which are plentiful in Wisconsin, the number one state in the production of canning peas and one of the leaders in potatoes. The various LPC concentrates have a balance of amino acids similar to that of alfalfa, a result Stahmann had expected. As far as scientists know, the balance of amino acids is much the same in all leaves. The press cakes made good silage, but for some reason, the cows didn't like the potato-vine silage. Wisconsin isn't the only area with large amounts of plant wastes. In the United States as a whole, Stahmann estimates, well over 21,000,000 *tons* of plant wastes containing 393,000 *tons* of protein are lost each year.

Plants of the brassica family, which includes species such as kale, collard greens, and mustard greens, may turn out to be among the better sources of leaf protein. The U.S. Department of Agriculture's Food Crops Utilization Laboratory in Weslaco, Texas, has pressed a plant juice from *Brassica carinata*, a native of Ethiopia, that has 50 to 60 percent protein and a lysine content of 6.6 to 6.8 percent, similar to that of alfalfa. Yields are very high and Texas A & M University, which is working with the USDA on the plant, estimates it might provide four to five hundred pounds of protein per acre. That would put *B. carinata* ahead of alfalfa. Unlike alfalfa, the Ethiopian immigrant—it was grown here for the first time in 1966—can be eaten by humans as a green leafy vegetable, the way it is utilized in its native land. The flavor is similar to that of collard greens, although milder. Thus far, it has been grown in only a few locations, principally Texas, but it has done well even in the North.

If you want to try raising some *B. carinata* for its fresh leaves (which, like other fresh leaves, have only 2 to 3 percent protein), the Texas Agricultural Experiment Station at Texas A & M released the plant under the name TAMU-TexSel. Write to the Foundation Seed Section, Department of Soil and Crop Sciences, Texas A & M University, College Station, TX 77840 for information.

Another multi-purpose plant being suggested for leaf protein is ramie (*Boehmeria nivea*), an Asian crop which is also grown in areas of Florida. Today ramie is planted only for the strong, durable fibers in its stem, which are used for clothing, string, and paper, but in recent years the protein content of its leaves has aroused the interest of scientists. According to the National Science Foundation, the protein in ramie leaves is similar in quality and quantity to that of alfalfa. Extracting the leaf protein does not affect the stems, making it possible to produce two salable products from one harvest. Ramie flourishes in warm, humid areas with heavy rainfall, making it a marginal crop in the United States, but it might supply a good source of leaf protein in tropical areas.

Unlike ramie, what may turn out to be the best plant of all for leaf protein is very much a part of the American scene: tobacco. Dr. T. C. Tso of the U.S. Department of Agriculture's Agricultural Research Service, Beltsville, Maryland, has developed a process known as Homogenized Leaf Curing (HLC), by which proteins are removed from tobacco leaves before they are cured for smoking tobacco. That means you can have your tobacco protein and smoke it, too. The key to the process lies in the fact that the compounds that enhance smoke and tobacco usability are different from those that contain protein, making it possible to separate them before curing. Two protein products result from HLC. The more valuable is Fraction-1 Protein, a pure, tasteless, odorless, and colorless protein that has a nutritional value, according to the

USDA, that is "comparable to human milk and surpasses that of soybeans." Suitable as animal and human food, it can be manufactured in a gel form that looks much like tofu.

Fraction-1 Protein, the major soluble protein in all green plants, is a familiar protein to biochemists, but thus far they have been able to crystallize it only from tobacco. The HLC process also makes it economically feasible to do so for the first time.

The other protein in tobacco, known as Fraction-2 Protein, is a mixture of many proteins that is also suitable as human and animal food. Dr. Tso estimates—"conservatively"—that about twenty to forty pounds per acre of both kinds of protein could be obtained as by-products of tobacco farming at current yield levels. He projects a worldwide yield of 12.5 billion pounds of tobacco by 1985 and 20 billion pounds by the year 2000. This translates into 750 million pounds of tobacco protein by 1985 and 1.2 billion pounds of protein by 2000. This amount of protein could meet the needs of thirty-three to sixty-three million people depending on which of several standards for protein consumption are used. Why not forget about smoking tobacco and just grow tobacco for protein? It's not economically feasible, according to Tso, who points out that tobacco plants contain only 12 to 17 percent protein. "But even half that 12 to 17 percent, which we can get as a by-product of normal production, can be very significant. We hope that in the year 2000, when the world will have six billion people to feed, we can make sensible use of the protein fractions and other useful products from tobacco that will otherwise literally be going up in smoke."

Other areas of the world have their own leading candidates for leaf protein. According to Dr. Richard G. Koegel of the University of Wisconsin's leaf protein team, the most promising plant for leaf protein in the tropics, where alfalfa cannot be grown, may turn out to be the most plentiful form of vegetation: trees. Dr. Howard Ream, another member of the team, suggests additional tropical plants for leaf protein: a variety of soybean that produces only forage, not beans; the koa haole of Hawaii, which has a very high yield of protein per acre; and plants such as the silverleaf and greenleaf which, like the tropical soybean, are legumes adapted to hot, wet areas. Cassava leaves, the foliage of the low-protein root crop eaten in many tropical areas, are another potential source of protein, in the estimation of Dr. Jack R. Harlan of the University of Illinois. Some wild plants may be suitable for leaf protein, too. One that has been suggested by several researchers is the water hyacinth, a prolific weed that impedes navigation in many warm areas of the world, including the southern United States.

Not all leaves, though, are suitable for leaf protein. Some yield poorly, others contain troublesome toxins, and the protein content of still others falls below the level at which an economic concentrate and press cake can be produced. Most of the low-protein group are grasses, some of which have only about 10 percent protein on a dry-weight basis, compared to 20 to 30 percent for most leaves. The University of Wisconsin's Dr. Neil Jorgenson, a dairy science professor and a coordinator of the leaf protein project, estimates that when all factors are taken into consideration, only about one-quarter of the world's leafy plants will qualify for protein extraction. But with some three hundred thousand plant species on earth, that still adds up to a big field of candidates. Thus far, most species have not even been screened for protein content. Somewhere, there may exist a leaf even better for protein extraction than alfalfa. Unless it's found, though, the modest-looking plant that makes most of the world's hay will continue to be man's best source of leaf protein.

7

COTTONBURGERS, PEANUT FLAKES, AND CORN-GERM CANDY

> Peas! Peas! Peas! Peas! Eating goober peas!
> Goodness, how delicious, eating goober peas!
>
> —CIVIL WAR MARCHING SONG

THE SOYBEAN TURNS UP IN SO MANY DIFFERENT FOODS TODAY that it sometimes seems as if it's the only source of plant protein available. It isn't the only one, however; it's simply the one on which the most work has been done. A large number of other sources of plant protein are grown in this country and some of them are almost as nutritious as soybeans. With processing, a few are even more nutritious. Four crops that are widely grown in the United States look like particularly good candidates for fulfilling some of our future protein needs: cotton, peanuts, corn, and oats. All are already used for food, of course. Cottonseed supplies a widely used cooking oil; peanuts are one of our favorite snacks and the source of a popular luncheon spread; corn on the cob is an American tradition; and oats are a staple breakfast food and cookie ingredient in many households.

With the possible exception of oats, however, none of these crops is eaten by humans for its protein content. Yet all, in one form or another, are excellent sources of protein.

A special kind of cottonseed flour has 70 percent protein, a defatted peanut has 43 percent protein, a flour made from corn germ not only has 25 percent protein but also a 5.9 percent lysine level, and an oat concentrate has up to 88 percent protein. A variety of tasty and nutritious products have been made in the laboratory with these plant protein sources, including meat extenders, breads, muffins, candy, shakes, and other beverages. One or two of these products are already available and others may be in your food store before long. By the end of the century, it is a good bet that all of them will be in our diet.

What happens to these sources of plant protein today? For the most part, they are fed to animals. Cottonseed meal, the press cake left after peanut oil is extracted, hominy feed incorporating corn germ, and oats are all fed exclusively or principally to animals in this country. Our livestock not only eat better than many people in developing countries, but often eat better than we do.

It's hard to pick which one of these animal feeds will be the first to become a significant source of protein in the human diet, but some scientists give the nod to the plant that sounds the most unlikely: cottonseed. We think of cotton primarily as a cloth, but according to nutrition researcher Margaret L. Harden of Texas Tech University in Lubbock, cotton actually produces more food for man and feed for livestock than it does cotton. For each 100 pounds of cotton fiber, the cotton plant gives some 170 pounds of cottonseed. Cottonseed is a small yellowish seed, the kernel of which contains 40 percent protein and 40 percent oil. The rich store of oil puts it in the group of plants known as oilseeds, to which the soybean and the peanut also belong. After processing, 170 pounds of cottonseed supplies about 27 to 30 pounds of oil and 72 to 82 pounds of defatted meal. In the United States, as in most cotton-growing countries, the oil, which is consumed by humans, is the most valuable product. Next in value comes the meal, which is eaten by cattle. A cheap product, cottonseed hulls, is also fed

PROTEIN AND OIL CONTENT
OF THE MAJOR OILSEEDS

OILSEED	PERCENT OIL	PERCENT PROTEIN
Cottonseed	16	21
Soybean	18	40
Peanut	45	27
Sunflower	27	19
Flaxseed	38	35

Source: Green, Texas A & M.

to cattle, although some are used for industrial purposes.

Worldwide, production of cotton is substantial—22 to 24 million metric tons. Figuring 20 percent of that as protein (the low-protein hull accounts for much of the weight of the whole seed, but has only half the protein content of the kernel), almost 4.8 million tons of cottonseed protein are produced each year. That's about 6 percent of the world's total supply of edible protein. In the United States, we produce some 4 to 5 million tons of cotton each year, making us one of the world's leading producers. Our big cotton-growing states are Texas, California, Louisiana, Mississippi, Georgia, South Carolina, and Oklahoma.

Much of the cottonseed utilization research in the United States is carried out at the U.S. Department of Agriculture's Southern Regional Research Laboratory in New Orleans. Like the USDA's three other regional centers, it concentrates on crops grown in the region where it is located. In the South, cotton is still king, although the SRRL also studies other southern crops, including peanuts. Most of their work is concerned with commercial products—cloth and oil in the case of cotton—but food development research has been increasing in the last decade. One of the results of this research is the emergence of a number of new cottonseed-based and peanut-based food items, all with high protein content.

I visited the Laboratory on a February day when it was almost as cold inside the sprawling complex of buildings near New Orleans's Lake Pontchartrain as it was outside. A federal institution, the Laboratory was being scrupulous about obeying President Carter's recent directive to keep thermostats at sixty-five degrees. In one of the high-ceilinged offices, Leah Berardi, a research chemist in Protein Products Research, shivered from time to time in her spring dress and lab coat as we talked about cottonseed. Berardi and Dr. John Paul Cherry, a research leader in Protein Products Research, spend most of their time working with cottonseed protein, which they predict will play an increasingly important role in the human diet. The small seed, according to Berardi, has certain advantages over soybeans.

"Cottonseed has a bland taste and it doesn't produce flatulence," she said.

I already knew, from reading material the SRRL had sent me, that cottonseed flour has a lysine level below that of soy flour—5.4 percent for cottonseed flour, 8.6 percent for soy flour—but that both are above the lysine level in the reference protein devised by the Food and Agricultural Organization of the United Nations. Cottonseed flour, however, comes a little closer than soy to the 2.2 FAO level for another amino acid, methionine. Cottonseed flour has 1.8 to 2 percent methionine, soy flour 1.1 to 1.8 percent. Beside methionine, the major amino acid deficiency of cottonseed flour is isoleucine.

Cottonseed foods designed for human use are not completely new. As early as 1876, a cottonseed flour containing up to 55 percent protein was produced in the United States. In the 1930's, a Texas firm came out with an improved cottonseed flour. Both of these early flours, however, had problems, most of them arising from the presence in the seed of a substance called gossypol. Gossypol is a compound that is located in dark-colored glands scattered throughout the seed kernel. It affects the digestion of nonruminants such as pigs, chickens, and man, and must be removed or inactivated before the cottonseed is fed to them. In the early cottonseed flours, gossypol was inactivated by heating the flour, which caused the gossypol to bind to the lysine. Unfortunately, binding gossypol to lysine lowered the amount of lysine available and made the cottonseed flour less nutritious. The presence of gossypol also made the flour dark.

Then, in the late 1950's, a "glandless" cottonseed without gossypol was developed by crossing various lines. The first glandless cottonseed was planted in 1966, and today it is grown extensively in Texas. Widespread use of glandless cottonseed, however, is still in the future. "It takes a long time to phase in something new like this," commented Cherry. "And we're a long way from glandless cottonseed here in Louisiana where there are so many insects. Gossypol protects the seeds from insects." The best bet in cottonseed foods for the near future, he believes, is a process developed at the Southern Regional Research Laboratory which produces a high-quality flour with very little gossypol. The process is called the Liquid Cyclone Process, or LCP for short.

In the LCP process, the gossypol is literally whirled out of a slurry composed of milled kernels and liquid onto the walls of a tank by a small ten-inch by three-inch metal device, the "cyclone." Together with the heavier components of the slurry that were whirled away with it, the gossypol drops to the bottom of the tank and then falls out. Meanwhile, the lighter residue, which contains most of the protein, stays in suspension in the tank and leaves by the top.

After several additional processing steps, there are two end products: a 35 percent protein meal with most of the gossypol, which is suitable for ruminants, and a 70 percent protein flour with very little gossypol, which is suitable for nonruminants. The flour, which is technically a concentrate because of its high protein content, is almost white and has little taste. Its PER ranges from 2.3 to 2.7, which compares very favorably with food technologists' favorite reference protein, milk casein, at 2.5. The flour is also the starting point for a process that results in another bland, white product, cottonseed isolate, which contains at least 90 percent protein.

To show me the difference in the various cottonseed products, Leah Berardi moved into her laboratory, where she had spread out a number of bottles. "This is our LCP flour," she said, picking up a bottle of whitish flour. It looked completely white to me until she put a bottle of glandless flour next to it. The glandless flour *was* white. "It's beautiful," said Berardi. "Now compare that with this." She brought over a bottle of dark flour, a sample of the commercial flour produced by binding gossypol to amino acids by heat. She completed the display by lining up bottles of two types of cottonseed protein isolate, both of which were a light yellowish color, similar to the LCP flour. "These isolates look the same but one of these has a 3.2 PER and the other has a 2.2 PER," she said. "The 3.2 one has more of one type of protein—the nonstorage type protein—than the other. We make two isolates because one performs better functionally for certain uses than the other." She picked up another bottle, this one full of seeds, and shook a few onto the table. "See those little green specks in the seeds? Those are the pigment glands with gossypol."

By putting my eye close to the tiny seed (dehulled cottonseeds are smaller than peas), I could just make out some minute green specks on the surface. The glandless kernels she showed me had no specks and were a uniform yellowish color.

LCP flour, glandless flour, the isolates, and even heat-treated glanded flour all have their special uses. To date, the most successful use of cottonseed protein in a food product is in a 25 percent protein (dry-matter basis) beverage called Incaparina, which was developed by American scientist Dr. Nevin Scrimshaw some years ago. It is marketed only in Central American countries, one of which, Guatemala, now has its own modern Incaparina processing plant, which produces ten tons a day. The Incaparina formula is composed of a cereal and an oilseed flour, plus vitamins. Since corn and cottonseed are readily available in Central America, the form Incaparina takes in Guatemala incorporates corn flour and cottonseed flour, both of which are milled right in the plant. Incaparina costs about one-third as much as milk, making it possible for even many low-income families to buy it.

In this land of plentiful and relatively low-cost milk, Incaparina would probably not find much of a market, but the New Orleans center has cooperated with researchers at several universities in devising cottonseed-based products that will appeal more to our tastes. One promising use is in meat extenders, where cottonseed protein appears to have many of the properties that make soy protein so useful. Texas A & M researchers have used various forms of cottonseed in meatballs, meat patties, and meat loaf with considerable success. Another use is in bakery products. Margaret L. Harden and S. P. Yang of Texas Tech University recently compared the results of putting 18.8 percent LCP flour, glandless flour with the same protein content, and glanded flour into loaves of yeast bread. The lysine content was raised from 2.4 percent in the unsupplemented bread to 3.8 percent in the LCP and glandless cottonseed-fortified breads (the figure is below that of the flour because baking diminished amino acid levels in this case). The total protein content of the supplemented loaves was 21 percent for the glandless loaf and 19 percent for the LCP loaf, both of which are almost double that of the 10 percent protein unsupplemented loaf. The glanded loaf? It raised the protein content too, but it didn't do the laboratory rats that ate it much good; they all died within five days. Rats that ate the LCP and glandless breads gained the same amount of weight.

But unlike laboratory rats, man does not live by bread nutrients alone. The cottonseed-fortified loaves prepared by the Texas researchers were somewhat heavy, dark, squat, and rough-textured, traits many Americans dislike in bread. Another research team put only half as much cottonseed flour —about 10 percent compared to 18.8 percent in the experiments described above—in bread and also reported decreased loaf volume. These results are not particularly surprising. Cottonseed flour, like soy flour, has no gluten, the substance in wheat that holds ingredients together and gives loaves, muffins, and other products their "spring." Thus decreasing the amount of wheat and, therefore, the amount of gluten in a wheat-based product, usually results in decreased volume. One way to get around the problem is to use cottonseed flour in bakery products where volume is not as big a factor as in yeast bread. The Harden-Yang team has worked out a number of cottonseed flour recipes for cookies, candy, quick breads, and crackers, some of which have been well accepted by taste panels. One of their most successful recipes is a cracker that has 40 percent protein.

Tortillas, the unleavened, pancakelike bread popular in Mexico, are another good candidate for cottonseed fortification. A group of Texas A & M researchers headed by Judy Green found that by adding 12 to 15 percent glandless cottonseed flour to tortillas, they could produce a version of the tortilla that a taste panel liked as much as the traditional corn tortilla. And they liked it much better than a soy-fortified tortilla. Cottonseed-fortified tortillas had 2.89 percent lysine with the 12 percent level of cottonseed, 3.20 percent with the 15 percent level. The unfortified corn tortilla had only 2.46 percent lysine.

The cottonseed product best suited to fortifying most leavened bakery products may turn out to be an isolate. In an experiment conducted at the Southern Regional Research Laboratory, Berardi and several colleagues discovered that adding 10 percent cottonseed flour to bread depressed loaf volume, but that adding 10 percent cottonseed isolate had little effect on volume. The protein content of the isolate-fortified bread was doubled. The 90 percent protein isolate also has been used successfully to fortify carbonated soft drinks. At the time I visited New Orleans, Berardi was in the process of patenting still another use for isolate—an imitation meat. She has devised a simple "heat and stir" method that turns cottonseed isolate into a textured product analogous to soybean "meat." She expects her product to be cheaper than spun soy meat.

"I think our cottonseed-extended hamburgers are better than soy-extended hamburgers," claimed Berardi. "They're tastier."

Cherry agreed. A local newspaper, he added, had come up with a name for the new product: Cottonburger.

But don't start looking for cottonburgers yet. Before all the good things that come from cottonseed can be incorporated into processed foods, a dependable supply of LCP or glandless flour, or both, is necessary. At the moment, comparatively little glandless flour is available and no plant in the United States is turning out LCP flour. A two-million-dollar LCP plant was set to begin operation in 1975 in Lubbock, Texas, but financial problems in the parent firm resulted in the completed facility closing before it had marketed its first batch of flour. There are rumors that it will reopen, but until it does, or a comparable facility goes into operation, nutritious cottonseed flour will remain a laboratory item.

There's just one exception to the present dearth of cottonseed products for humans: Cot-N-nuts. The Rogers Delinted Cottonseed Company of Waco, Texas, plans to begin selling a small, nutlike product made from the seed of glandless cottonseed to consumers under the catchy name Cot-N-nuts by 1978. The small, yellowish kernels have 37 to 38 percent protein. They look and taste much like chopped nuts, but sell for less than any nuts except peanuts. Cot-N-nuts are not available to consumers as I write this, but Pillsbury, the big flour-making firm, had some on hand and sent me a few sample packages. The kernels are a bit small for eating from the hand (I kept dropping them behind the sofa cushions), but they tasted very nutty in some cookies I made. No chopping was necessary, making their small size an advantage in baking.

Peanut foods, a subject to which another group of SRRL researchers is devoting its efforts, have an entirely different set of problems in gaining acceptance here. The unusual legume—the peanut belongs to the same family as the bean and the pea—is well liked by Americans, but only in certain traditional forms: peanut butter, peanut candy, and a variety of roasted and boiled peanuts eaten as a snack. Of the almost four billion pounds of peanuts produced annually in the United States, a little over half goes into peanut butter. The other food uses, in order of importance, are salted peanuts, peanuts used in candy, and roasted peanuts. Each of the three major types of peanut grown in the United States has its own particular use or uses. Virginia, the biggest, is usually eaten salted or roasted as a snack; runner, a smaller peanut, almost always ends up in peanut butter; and Spanish, a small, round nut, is used for all types of peanut products. In addition to peanut foods, about one-quarter of our peanut crop is pressed for oil. The defatted press cake is fed to animals.

Most of us probably eat peanuts because we like the way they taste, but peanuts are a nutritious food, not just a snack. They have about 25 percent protein, a little above average for a legume, and although slightly deficient in lysine, methionine, and threonine, they can easily be supplemented. Try a glass of milk (soy or dairy) with your peanut butter sandwich since wheat bread, like peanuts, is low in lysine. Compared to most other legumes, peanuts have one big nutritional advantage: They produce little or no flatulence. A team of University of California researchers led by Dr. Doris Calloway tested a wide variety of legume foods on human subjects recently, including a generous amount (about six ounces) of roasted peanuts. Their verdict: "Peanuts were absolutely non-flatulent."

The peanut is grown extensively in southern and southwestern states, Georgia being the biggest producer with about 45 percent of the total crop. In this country, the peanut has a curious history. A native of South America, it was discovered by the Spanish in the fifteenth century and shipped by them to Africa, where it found the climate to its liking. One of the popular names for peanuts, "goober" or "goober pea," is thought to be of African origin. Brought to this country by African slaves, the peanut again adapted to its new home.

By the Civil War, it was widely planted in the South. But peanuts really took off commercially after the war, thanks partly to the efforts of George Washington Carver, the black scientist, who was a tireless promoter of peanuts. Peanuts got another big boost in the 1890's, when a St. Louis physician invented peanut butter. Peanut butter sales zoomed in the 1940's, when additives came into use to prevent oil separation. Health food store and homemade peanut butter still have oil separation.

If peanuts are already so popular in the United States, why do we need additional peanut products?

The answer is that popular as peanuts are for some uses, they do not supply a significant amount of the protein in our diet. The peanut is much more widely used as a protein source in some other countries. In Indonesia, the nutritious press cake left over after the oil is removed is fermented into a kind of tempeh called *ontjom*. The Chinese sprinkle peanuts into a number of dishes in which meat is used in small amounts, thus augmenting the protein content of the meal considerably. West African dishes incorporating peanuts are usually meatless, for example, "peanut stew" that turns up in a number of countries. Since this country has an ample supply of the nutritious peanut—in fact, it is a little *too* ample (more on this later)—researchers in the Southern Regional Research Laboratory and research groups affiliated with private firms would like to change the image of the peanut from that of a snack to that of a staple.

To do it, they are developing a number of peanut products, some of them in conjunction with researchers at various southern universities. The products include a peanut-flake breakfast food, flavored peanut spreads, a high-energy "space-food stick" the size and shape of a cigar, and a defatted peanut flour. The flour is an improved version of defatted peanut meal, the high-protein animal food left after peanut oil is extracted. Taste panels gave some of the flour-based products high marks, particularly a muffin, a yeast bread, and an ice cream, the last with a rich peanuty taste. Not all of the products taste peanuty; the goal in some uses is to develop a bland product with the protein content but not the taste of peanuts. The New Orleans center's latest peanut product is a bland flour suitable for fortifying bakery goods or as a base for a drink.

"Taste it," urged Joseph Pominski, a chemical engineer who helped develop the product, offering me a jar containing a white powdery substance. I spooned out some.

"No taste," I said in surprise.

"That's right," said Pominski. "It's bland and cereallike. You can add any flavor you want. It has 60 percent protein, so you could put 5 to 15 percent in bakery goods and increase the protein content substantially. You also can use it to extend meat, the way soybean extenders are used. We've developed two different kinds of peanut milk drink from this flour, too." By subjecting the peanut flour to further processing, he added, peanut concentrates with about 70 percent protein and peanut isolates with about 90 percent protein can be produced. The SRRL recently installed a new facility with equipment adapted from the dairy industry to produce two different kinds of isolates from both peanuts and cottonseed.

The current emphasis at the New Orleans center is on making peanuts into a staple food, but the peanut product on which Pominski and his colleague, James J. Spadaro, a research leader, have done the most research in the past is more of a snack. It is a form of defatted peanut which, Spadaro told me, has less than half the calories of ordinary peanuts. A study that Spadaro, Pominski, and other researchers at the SRRL carried out shows that forty-one full-fat peanut halves have the same caloric value as ninety-five defatted peanut halves. At the same time, defatted peanuts have considerably more protein than full-fat peanuts—43.2 percent compared to 26.0 percent. This is good news for people who can't stop eating peanuts once they start (a group that includes myself), particularly since defatted peanuts are very good. I ate several handfuls while Spadaro and Pominski were talking and found they tasted much like dry-roasted peanuts. To make defatted peanuts, the nuts are pressed to remove the oil, then expanded in boiling water to their original shape. Finally, they are dried and roasted.

Although defatted peanuts have been around for a decade now, they've just come on the market in the Northeast under the name Peanut Peanuts in some supermarket chains. You can also buy them on the West Coast under the Bell Brand Foods label. You may be eating defatted peanuts without knowing it, though. The Carnation Company uses them in Breakfast Bars, and the Seabrook Blanching Company of Edenton, North Carolina, has developed artificial pecans, black walnuts, butter pecans, and toasted almonds, all based on the defatted peanut. Spadaro and Pominski showed me samples, and they look surprisingly like the product they are supposed to resemble. Seabrook, which stamps its mailing envelopes, "We work for peanuts," sells only to commercial users, so the walnuts or pecans you eat in bakery products or ice cream may really be defatted peanuts. The reason for the substitution is cost. Peanuts, while not cheap, still sell for less than other nuts.

The relatively high price of peanuts is a major drawback to their wider use as a vegetable protein source, though. Peanuts cost more than such other vegetable proteins as soybeans and cottonseed. The high price of peanuts is primarily the result of a controversial subsidy program for the peanut farmer that began back in the Depression era. The subsidy program assures the peanut farmer a certain sum for his crop and also guarantees that all of it will be purchased by the federal government if necessary. Year after year, huge peanut surpluses occur—the average surplus is about 1.2 billion pounds, according to *The Wall Street Journal*—and all of those surplus peanuts are purchased by the government at the subsidy price, which changes from year to year. What does the government do with a billion-plus pounds of peanuts? It sells them to the highest bidder. The peanut subsidy is under fire these days, but no one expects it to vanish completely.

In spite of subsidies and the popularity of peanuts as a snack, peanuts are too good a protein product to pass up. A few peanut foods are slowly emerging from the laboratory into processed foods, if not directly into the consumer's shopping bag. At least two peanut processors—Gold Kist and Seabrook—now offer peanut flours to commercial customers. Seabrook's has around 40 percent protein. Another peanut product, peanut flakes, looks as though it may be on the verge of commercialization. The latter, the creation of Clemson University professor J. H. Mitchell, Jr., a peanut researcher with many years of experience in the field, is a bland, white, instantly rehydratable product that can be mixed with a wide range of foods to extend volume and raise protein content. The flakes have about 33 percent protein. Some of the foods in which Mitchell has used his flakes successfully are spreads, lunch meat, candy, and eggs. A corporation to handle licensing and distribution has been set up, so peanut flakes may soon make the leap from a concept to a commercial product.

Mitchell sent me samples of a number of his peanut-flake products, each one consisting of a powdery substance of a beige or brownish hue in a plastic envelope. I followed the directions on the envelope and produced (with the addition, in some cases, of other ingredients) an omelet, scrambled eggs, pancakes, brownies, several kinds of meat-flavored cracker spreads, salmon croquettes, and salmon paté. They were extremely easy to make and every one, without exception, tasted good, although not at all peanuty. The flavor is definitely superior to that of soy, making peanut flakes a more acceptable extender. My home taste panel of two agreed on the salmon croquettes and ham cracker spread as their favorites. Peanut-flake products keep without refrigeration and look like a particularly good choice for boating and camping trips when refrigeration is unavailable.

Several other sources of vegetable protein based on staple crops grown in the United States are being studied at the Northern Regional Research Laboratory in Peoria. One of them is corn, our number one crop in acreage, production, and value. We already eat corn, of course, but not very much of it and not in the nutritious form in which it has been developed at the NRRL. People eat only about 20 percent of the corn grown in the United States; the rest goes to animals. The corn product in which the Peoria center is particularly interested today is corn germ flour, which has a protein content similar to that of legumes, but an amino acid balance closer to that of the egg. The corn germ is the small inner portion of the kernel from which the new plant grows should the kernel undergo germination. The germ makes up about 11.5 percent of the weight of the whole kernel. After the oil is removed, it has some 25 percent protein, a PER of about 2.19, a BV that hovers around 70, and 5.9 percent lysine as compared to 2.5 in the whole kernel. Corn germ is somewhat low in the amino acids methionine, tryptophan, and isoleucine, but its amino acid pattern puts it among the best sources of plant protein.

ESSENTIAL AMINO ACID COMPOSITION OF
WHEAT GERM AND DEFATTED CORN GERM FLOUR

AMINO ACID	CORN GERM (%)	WHEAT GERM (%)
Lysine	5.07	5.71
Threonine	3.37	3.47
Cystine	1.88	0.86
Valine	4.38	3.60
Methionine	1.26	1.29
Isoleucine	2.53	2.54
Leucine	5.84	5.17
Tyrosine	2.59	2.29
Phenylalanine	3.17	3.02

Protein content: corn germ flour, 24.29%; wheat germ, 36.21%, dry basis.

Source: Tsen, Mojibian and Inglett, 1974.

And there's a lot of it around, too. Corn germ, as Dr. Charles W. Blessin, research leader in the Cereal Science and Food Laboratory, explained to me, is a by-product of a process called dry milling, through which some 120 million bushels of corn pass each year on their way to becoming oil, meal, grits, and other products. Corn germ makes up about 12 million bushels of the total.

So where has corn germ been all our lives? Like cottonseed meal, peanut press cake, soybean meal, and some other high-protein vegetables that could readily be consumed by man, corn germ is fed to animals. Corn germ is part of what is known as hominy feed, a popular feed for cattle. Hominy feed has about 12 to 15 percent protein, largely because of its germ content, compared to about 9 to 10 percent protein in corn products such as flour, meal, and grits. All three of these products come from the low-protein endosperm, which makes up most of the kernel. One reason why cows get the most nutritious part of the corn kernel is that, with the exception of the germ, it contains a high amount of fiber, which nonruminants like ourselves find hard to digest. Corn germ, however, has only about 4 percent fiber. Since nutritious corn germ is digestible by man, the NRRL researchers argue, man should eat it, not the cow, which can thrive on other, less nutritious plants.

"Corn germ is just too good a protein to put in animal feed," Roger Eisenhauer, the NRRL's director of public relations, tells visitors.

Eisenhauer keeps a big bottle of corn germ flour on his desk. A bland, light-colored substance, the flour goes well with a number of products. Research carried out at Peoria shows that it extends ground beef while lowering protein content only slightly, keeps water-oil emulsions together as well as soy, and raises the low lysine levels in baked goods without decreasing volume if certain substances are added to the baking formula (see p. 51). Unlike soy, corn germ has no objectionable taste, making it possible to use it in larger amounts. The NRRL researchers even put some in fudge where, according to Eisenhauer, it tasted delicious. Mrs. Eisenhauer also has used it successfully in pancakes, waffles, and chocolate chip cookies. Another use for the versatile product is as a sort of instant supplement that can be sprinkled over foods such as cereals the way wheat germ is today. Tests at Peoria show corn germ has slightly higher percentages of most essential amino acids than wheat germ, although its total protein content is lower, 25 percent compared to about 36 percent.

But diverting corn germ from cows to people is not any easier than diverting peanuts or cottonseed. The availability of corn germ flour, like that of most other alternative sources of protein, depends on production by commercial food processors. Corn germ is a by-product of dry milling, a process in which the various components of the corn kernel, including the germ, are separated before oil removal. Corn oil is a big consumer item in this country and the most valuable product produced by dry milling. There are many dry milling operations in the United States but only one—the Lauhoff Milling Company of Danville, Illinois—has the equipment to remove oil from the germ in a way that makes the germ suitable for human use. At this writing, Lauhoff was not preparing corn germ flour on a commercial basis and the nutritious product remains strictly a cow's delight.

Another nutritious plant that is fed principally to animals in this country is oats, although a small part of our oats (about 5 percent) does go into human food products. Yet scientists have known for years that oats have the highest protein content among the grains, ranging from about 13 percent to 18 percent and more in new high-protein varieties (see Chapter 5). Too good for livestock? Two scientists at the NRRL think so. Over the past few years, research chemists Y. Victor Wu and James E. Cluskey have produced a number of high-protein food additives from dehulled oats (or groats, as they are known), by two methods. One, the classic method, involves concentrating the protein by removing starches and sugars. The other, which has not been successful with other grains, involves dividing the groats into their components, or "fractions." The oat foods produced by these two methods range from 60 to 88 percent protein, with a very favorable balance of amino acids. Lysine, in which grains are low, is adequate or high in almost every case. Oat protein, according to Wu and Cluskey, has practically no taste and has functional properties that match those of soy in some respects, making it a promising nutritional additive.

"We put it in milklike and citrus-type beverages with good results," Cluskey told me. The favorite of the taste panel, he added, was a chocolate drink with about 4 percent oat protein.

If oats come out of the feed trough, can corn germ, peanut flour, and cottonseed be far behind? Move over, Bossy.

8

STRANGE PROTEIN

"If you plant these beans over-night, by morning they will have grown up right into the very sky."

—*Jack and the Beanstalk*

THERE ARE, IT IS ESTIMATED, SOME THREE HUNDRED THOUsand plant species in the world, but man has used only three thousand or so for food throughout the ages. A much smaller number—perhaps one hundred fifty—have been cultivated on a commercial scale. Today, the number of major plant food sources, the plants on which most of the world's population depends for its proteins and carbohydrates, is surprisingly small. Some scientists estimate it at about twenty. Others put it at no more than fifteen. The short list includes the cereals, a few root crops such as the potato, a number of legumes, and a small assortment of carbohydrate sources such as sugarcane. And not only do we rely on a mere handful of crops for most of our food, but those crops rest on a very narrow genetic base as a result of selection for yields and other agronomically advantageous characteristics. In the United States, for instance, about 40 percent of all our hard red winter wheat, our bread wheat, is planted to two varieties and their derivatives. A similar situation prevails with regard to wheat and rice in some developing countries as an outgrowth of the Green Revolution, which stressed the planting of a few high-yield varieties. Many of the plants once grown in various countries are irretrievably lost as a result of concentration on a few species. According to one estimate, 66 percent of the oats, 90 percent of the soybeans, and 98 percent of the clovers introduced into this country have been lost.

The shrinking number of crops and their narrow genetic base worries many agricultural scientists. What, they ask, will happen if some new insect pest or one of the forecasted climate changes adversely affects a widely planted staple crop? The answers involve words like "catastrophe" and "disaster."

A partial solution to the problem exists and steps are already being taken toward it. One involves searching out neglected food species and wild relatives of domesticated species, collecting their seeds, and depositing them in banks. In 1958, the United States established the National Seed Storage Laboratory in Fort Collins, Colorado, a large facility that has some one hundred thousand seeds on hand today and has room for many more (see Chapter 5). Other new seed banks are in existence in a number of Western nations. Seed from these banks can be used to breed desirable new characteristics into our staple crops in the future. Seed banks also serve another purpose. They provide a source of alternatives to our staple crops, alternatives which may become staples in the event of an agricultural disaster striking one of our small list of staples.

But we may not have to wait for an agricultural disaster before getting a taste of some of the alternative plant species. Some of them look too good to keep in a seed bank. In 1974, a scientific panel of the National Academy of Sciences, a Washington-based private organization which examines scientific subjects at the request of the federal government, undertook a survey of underexploited tropical plants. It came up with thirty-six species, some rich in protein, others in carbohydrates. A few are sources of industrial products. Some of the most promising, the Academy points out in its 1975 book *Underexploited Tropical Plants with Promising Economic Value*, are once-popular indigenous species that were largely abandoned at the advent of the colonial era in favor of crops of interest to the European consumer. Others simply never received attention beyond a restricted area.

The most outstanding species among the thirty-six, and the only one on which the Academy has issued a separate report, falls into the latter category. *Psophocarpus tetragonolobus,* a legume called the winged bean in English, is known only in New Guinea and a few areas of Southeast Asia. Although it's sold in some markets in small quantities, it's principally a home garden crop. Nutritionally speaking, the winged bean

ESSENTIAL AMINO ACID COMPOSITION OF WINGED BEAN COMPARED TO SOYBEAN

AMINO ACID	WINGED BEAN	SOYBEAN
Cystine	1.6–2.6	1.2
Lysine	7.4–8.0	6.6
Threonine	4.3–4.5	3.9
Valine	4.9–5.7	5.2
Methionine	1.2	1.1
Isoleucine	4.9–5.1	5.8
Leucine	8.6	7.6
Tyrosine	3.2	3.2
Phenylalanine	4.8–5.8	4.8
Tryptophan	1.0	1.2

Source: Cerny, 1971, and Kapsiotis, 1968, *British Journal of Nutrition,* Cambridge University Press.

sounds almost too good to be true. The seed has 34 to 37 percent protein and 18 percent oil, about the same percentages as soybeans, but the obscure legume also has a tuberous root, the protein content of which ranges from 20 to 24 percent, dry weight. No other known tuber even comes close to that protein content. The Protein Efficiency Ratio (PER) of the seed is 2.14, high for a bean. Soybeans average about 2.3. And that's not the end of the winged bean's nutritional virtues.

On a fresh-weight basis, the leaves have 5.7 to 15 percent protein, the flowers 5.6 percent protein, and the seed pods 1.9 to 2.9 percent protein. And the whole plant is edible— seeds, tubers, leaves, and flowers. Seeds must be cooked to inactivate an antitrypsin factor like that in many beans, but the rest of the plant can be and is eaten fresh as well as cooked. The flavor lacks any bitterness and even people unfamiliar with the winged bean take to it with enthusiasm after a bite or two.

"Believe me, the plant tastes good," says Dr. Theodore Hymowitz of the University of Illinois, who grows the plant in the university greenhouses. "The flowers taste like mushrooms fried in oil. You can eat the whole thing like an ice cream cone."

The winged bean's advantages aren't solely nutritional. It is a vigorous plant, so much so that when staked it grows tall enough within a few months to require a ladder to pick the highest pods. Twelve-foot plants are not unusual. In New Guinea, a plant will bear fifty to sixty pods from two to fourteen inches long, each containing up to twenty seeds. With no large commercial acreage in existence, yield figures must be extrapolated from experimental plots, but in southern Florida seed yields of up to one ton per acre have recently been achieved. This would put the winged bean in the same league as the soybean and peanut. The plant seems to be remarkably resistant to insect pests and it needs even less nitrogen than most species on newly cleared land, which is deficient in soil nutrients.

When I was reading all this in a booklet called *The Winged Bean: A High-Protein Crop for the Tropics*, I kept imagining how the winged bean would appear in a typical American seed catalog.

FANTASTIC NEW BEAN
Two high-protein crops with one plant!
Sixty pods to a plant!
Eat it all: pods, beans, leaves, flowers, roots!
This incredible new bean from the lush island of New Guinea

provides a bean as nutritious as the soybean but far more delicious, as well as the world's highest-protein root. And you can eat the high-protein pods, leaves, and flowers, too! Resistant to insect pests, it grows to twelve feet high and bears up to sixty pods per plant. A small plot can supply most of the protein needs of the average family during the growing season.

Unfortunately, an advertisement of this nature is strictly fantasy, at least as far as this country is concerned. The winged bean is a species of the humid tropics, although it grows well in the dry tropics under irrigation. In our country, that means it would probably be restricted to such areas as Hawaii, Florida, California, and the territory of Puerto Rico. The winged bean's expansion to other areas of the tropics, however, looks very feasible. "With research, the winged bean can become a significant crop in the humid tropics," says the National Academy of Sciences. It recommends a research effort to develop, among other things, high-yielding and sturdy dwarf varieties amenable to commercial harvesting. Research, the Academy suggests, may even give us a variety of winged bean that would flourish outside the tropics. "Many advances in the genetic improvement of the soybean could be applied to the winged bean," the NAS notes. "It is possible that the winged bean may be as significant for world agriculture in the future."

One indication that the winged bean may, indeed, be able to survive outside the tropics is a successful field trial conducted right near my own home in northern New Jersey. T. M. Mozell of Somerset County, N.J., planted winged bean seeds in the spring that grew into plants over 2.7 meters tall (almost nine feet). They survived a tropical storm and a hurricane but were killed by heavy frost in the fall.

On the chance that the winged bean is the soybean of the future, I stopped at the University of Illinois in the heart of soybean country one hot day in August to look at some New Guinea specimens growing under Dr. Theodore Hymowitz' care. Hymowitz himself had hoped to be there, but at the last moment he had to fly to Hawaii to participate in a conference on the winged bean, which is receiving an encouraging amount of scientific attention at the moment. I talked, instead, to his research assistant, Dr. Christine Newell, a young woman who is a specialist in cytology, the branch of biology dealing with the cell. Her own particular interest in the winged bean is the determination of its chromosome count, which, like most of the basic facts about the plant, is still unknown. After she showed me some of the pea-sized beans, which range in color from cream to black, we walked to the greenhouse area near the agronomy building. The beans, Newell explained in her pleasant British accent, had been moved outside, the August heat inside the non-air-conditioned buildings being too much even for a species from the tropics.

"Well, there they are," she said when we reached a paved area between two of the greenhouses.

For a protein wonder, the plants looked rather unremarkable, although very healthy. They were extremely leafy and about four or five feet high. Each one twined around a stake thrust into a large plastic garbage can. They had been cut back to stumps in June, Newell said, then moved outside. In the fall, they would go back in again, where they would get much taller. "If we didn't cut them back, they'd shoot right up to the ceiling. I left one alone and it did grow up to the ceiling." At this stage, the plants were just coming into bloom and Newell searched in the heavy foliage for a flower, finally coming up with a rather small blue one. Winged bean flowers also come in white and purple. "It looks like it's going to set a pod," she noted. "The pods grow out of the flower's pistil, this little structure here." In New Guinea, a plant bears up to sixty pods, but in Illinois, they are lucky to get fifteen to twenty per plant. In areas where the tubers are a popular food item, Newell added, the blossoms are pinched off before they set a pod, a practice that seems to enlarge the roots.

Winged beans have to overcome two major problems before they can be cultivated on a large commercial scale, according to Newell. In the wild, the plant rambles along the ground, but under cultivation methods used in New Guinea, it is staked to increase yields. Staking, however, makes mechanical harvesting impossible. "To make the plant stand up by itself, it will have to be dwarfed," Newell said. Another problem lies in the fact that the pods ripen at different times. Neither trait is unusual in a wild plant; soybeans, for instance, were a rambling ground plant before genetic manipulation produced the short, upright plant familiar today. Breeding, believes Newell, can produce changes in the winged bean analogous to those that have been made in the soybean.

After I returned home, Hymowitz sent me a new paper on the winged bean he had written with Dr. John Boyd of San Luis Obispo College (Calif.). It suggests a New Guinea origin for the species, although the other five members of the family to which the winged bean belongs hail from Africa. One reason why New Guinea looks like a plausible spot for its origin is that people living on the large tropical island in the South Pacific use more parts of the winged bean than are used in any other area. New Guineans eat the pods, seeds, flowers, leaves, and roots, a range of uses which suggests long familiarity with the plant's nutritive value. One favorite New Guinea recipe involves wrapping the tubers in banana leaves and roasting them on an open fire. The only other place where the tubers are eaten is on some of the other South Pacific islands and in Burma. According to Hymowitz and Boyd, the roots are firm and slightly sweet, rather like an apple. They can also be eaten raw. The most peculiar use of the winged bean found by the researchers occurs in Java. There the swollen parts of young plants that are parasitized by a fungus are steamed and eaten as a delicacy. There's no end, it seems, to the uses of the winged bean.

For the near future, though, the versatile winged bean will remain a tropical species unavailable to most Americans. Several other high-protein plants discussed in the National Academy of Sciences' book *Underexploited Tropical Plants with Promising Economic Value* look like much better candidates for our own use. Some were eaten by the American

Indians and several others have been grown here experimentally and commercially with some success. Among the half dozen or so plant protein sources that fall into this category, one looks particularly promising: the buffalo gourd. The buffalo gourd, *Cucurbita foetidissima*, is a wild plant that is indigenous to the southwestern United States and northern Mexico. If you've ever driven through the Southwest and noticed a low, vinelike plant with large, heart-shaped leaves the size of a man's hand, you were probably looking at a wild buffalo gourd. An extremely drought-resistant species, it can survive a year without water because of the water retention qualities of its roots. Some plants have been dug up with huge branched roots as big as a man. One branch of such a root weighed one hundred pounds. The root is fascinating from a botanical point of view but not particularly nutritious, being low in protein and containing bitter-tasting substances called glycosides. Soaking in salt water removes the bitterness, but the real nutritional value lies elsewhere. An average plant bears up to one hundred round yellow gourds the size of tennis balls. Inside each ball are some three hundred flat, oval-shaped seeds about one-half inch long.

Analyses of the seeds show they have over 30 percent protein and about the same percentage of oil, figures that put them well ahead of cottonseed, but behind the soybean in protein content. Some plants have given seed with about 35 percent protein, comparable to soybeans. Like the protein of most plant sources, buffalo gourd protein is low in the sulphur amino acids methionine and cystine and borderline in lysine. Even the leaves and stems are high in protein, 17 and 11 percent, respectively. Buffalo gourd meal made from seeds after the oil has been removed has about 50 percent protein, a higher content than any oilseed meal including soybean meal. The oil is of high quality, with some 60 percent polyunsaturated fats. In the latest experimental plots, buffalo gourds yielded between two thousand and three thousand pounds of seeds per acre, the equivalent of between thirty and forty-five bushels of soybeans per acre. The average

United States soybean yield is almost thirty bushels per acre.

The obscure wild plant from the southwestern deserts has still another merit. When the seed is mature, the skin of the gourd becomes thin and opens easily, spilling out the seed. This feature makes it easy to harvest the seed with existing farm equipment, a big attraction in commercialization of new species.

With all the buffalo gourd has going for it, you'd think it would be a cash crop in dry areas of the world, but until recently its attractions were one of the better-kept secrets of agronomy. As far as is known, the only people to use it as food were the American Indians, who ground up seed from wild plants into a kind of mush and also ate seeds whole. The neglect seems even more curious when you consider that many of the buffalo gourd's relatives are popular food crops. The buffalo gourd belongs to a family called the cucurbits (pronounced Q-cur'-bit), which includes all the squashes and melons. Cucurbit fruits are popular throughout the world, but other parts of the cucurbits also are used as food. Cucurbit seeds are eaten in many parts of the world and cucurbit oil is a popular edible oil in Eastern Europe.

The first scientist to realize the potential importance of the obscure buffalo gourd was Dr. Lawrence C. Curtis, a squash expert who recently retired from the University of Georgia. A native of Arizona, Curtis had seen buffalo gourds growing wild on the desert as a boy. In 1946, he wrote an article on their value as a protein and oil source in the *Chemurgic Digest*, a scientific journal. In those days of plentiful and cheap soybeans, the article did not arouse much interest among agronomists and Curtis turned his attention to other cucurbits, developing, among other things, an improved variety of summer squash. By the 1960's, however, new sources of plant protein were actively being sought and Curtis' buffalo gourd took on new importance. The Ford Foundation sent him to Lebanon to raise the buffalo gourd in that area, which has a climate much like our Southwest. By 1974, Curtis was getting yields of close to three thousand pounds per acre in Lebanon. The Lebanese civil war ended the project, at least temporarily, and Curtis returned to the United States to work with a group of researchers at the University of Arizona who had begun growing buffalo gourd in its home territory.

Some of the figures the five-man Arizona team (besides Curtis, they are William P. Bemis, Charles W. Weber, J. W. Berry, and J. W. Nelson) have published in the last few years indicate the buffalo gourd may be an even better source of protein than its high protein content indicates. In a feeding trial with mice, for instance, the PER was 2.05; with rats, it was 1.66. Milk casein, the widely used reference protein, is 2.50. The BV of buffalo gourd protein is about 55, a figure that puts it below other oilseeds but ahead of most beans. Chicks that ate up to 23.5 percent buffalo gourd seed in their diet gained a little less weight than chicks on a control diet, but the difference was not significant.

The buffalo gourd has a few drawbacks as a source of protein for humans. As mentioned earlier, it's low in methionine and cystine and marginal in lysine. Also, the seeds are high in fiber, 6.9 percent in the hull-free seed and 11.8 percent in the defatted hull-free seed, a content that would restrict their consumption by humans. The meal, which has most of the oil removed, has about 10 percent fiber. Soybean meal, by contrast, has only 3 percent fiber. Another problem is the dark color of the oil; in this country, light-colored oils are considered desirable. Commercial processing of oilseeds is built on a two-step system in which the oil is first removed, leaving an oil-free residue that can be further processed for human consumption. If the oil is unusable, the whole process becomes uneconomical. Buffalo gourd oil can be bleached to a lighter color, but it is "resistant" to bleaching, according to researchers.

But there's no alternative or even traditional source of protein without a problem and some experts think the buffalo gourd looks like a winner. William P. Bemis of the University of Arizona predicts the gourd could eventually become the world's greatest oilseed crop. He predicts "extensive production" within six years.

The American Indians also recognized the nutritive value of several other plants discussed in the NAS's *Underexploited Tropical Plants*. Tribes in both the Northern and Southern Hemispheres ate a grain called quinua (*Chenopodium quinoa* Willd), together with some of its relatives among a family called chenopods. Another popular Indian grain belonged to the amaranth family. Both families are ancient. Quinua seeds dating back two thousand years have been excavated at an archaeological site in Argentina. Massive caches of amaranth seeds have been found associated with such diverse cultures as the Hopewellian in Illinois, the Basketmaker in Colorado, and agricultural Indians in Mexico. A cultivated grain amaranth may have persisted in this country until recently, according to Dr. Jonathan D. Sauer of the University of California. About 1890, the custodian of Casa Grande National Monument in Arizona was told by members of the Pima tribe that a hundred years earlier, their forebears had eaten a "small round seed which they ground and boiled as mush." This description fits the amaranth and seeds of the plant have been found nearby, at Tonto Basin, where agriculture was practiced by Indian tribes.

Dr. John R. K. Robson, of the Medical University of South Carolina in Charleston, furnished me with a list of the ways in which wild amaranths and chenopods were used by North American Indians in historical times. The Navahos and Shoshone ground the seeds of amaranths into a flour, which they baked into loaves of bread. Several other tribes ground the seeds into meal and ate the young leaves fresh or boiled. Chenopods were more often eaten as greens than grain. The Hopi picked young greens in the spring, put them in a pot, and boiled them for several hours (no concern about vitamins at that time, of course), then served them with salt, fat or butter, and corn dumplings. The Navahos, Pueblos, Diggers, and Utah, among other southern tribes, also ate chenopods as raw or boiled greens. When the seeds were used, they were ground into flour and made into bread or mush, sometimes with another grain, such as corn. One agricultural researcher who observed the Pima Indians in the late nineteenth century furnished their recipe for quinua mush: "Dry and grind seeds. Mix one to two handfuls of meal to one pint of boiling water. Add one teaspoon of salt. Cook until done."

Farther south, in Central and South America, domesticated amaranths and quinua were widely cultivated at the time of the Spanish conquest. Montezuma collected a tribute of some two hundred thousand bushels of grain, including *Amaranthus*

ESSENTIAL AMINO ACID COMPOSITION (%)
OF QUINUA AND WHEAT

AMINO ACID	QUINUA	WHEAT
Histidine	2.7	2.1
Lysine	6.6	2.7
Tryptophan	1.1	1.2
Phenylalanine	3.5	5.1
Methionine	2.4	2.5
Threonine	4.8	3.5
Leucine	7.1	7.0
Isoleucine	6.4	4.0
Valine	4.0	4.3

Source: White *et al.*, 1955.

hypochondriacus, from his provinces every year. One of the important uses of amaranth grain among the Aztecs was as the basis for a paste called zoale, which was made into the shapes of religious figures. The Spaniards forbade the use of *A. hypochondriacus* because it was associated with "pagan" religious observances; the other species and quinua were largely supplanted by grains like corn and barley. Ironically, these grains are much less nutritious than the ones they replaced. The amaranths have 15 to 18.5 percent protein;

quinua ranges from 12 to 22 percent. A semidomesticated relative of quinua that is grown in Peru and Bolivia is said to have even more protein. These levels put the amaranths and chenopods well ahead of any of our major grains. The lysine levels of these neglected plants are even more impressive. One species of amaranth has been shown to have 6.2 percent lysine; quinua has 6.7 percent. These are the highest lysine levels known for any food grains, which are usually low in lysine. Wheat has less than 3 percent lysine.

Although amaranths and quinua do not enjoy the popularity they once had, they are still cultivated in Central and South America, as well as a few other areas. Different species of amaranths are grown in areas of Argentina, Peru, Bolivia, and Mexico. The grain is parched and milled and the flour made into a kind of pancake or into a gruel or a drink. Young leaves are eaten as a green vegetable, which also is reputed to be high in protein. The plant also is found in India, China, and Nepal, areas to which it was transported by ship after the Spanish conquest. Quinua, a plant that flourishes at high elevations, is grown in the Andean highlands of Bolivia, Chile, Ecuador, Peru, and Argentina. Dr. Hugh D. Wilson of the University of Wyoming, who wrote his doctoral dissertation on the domesticated chenopods, reports quinua is grown today primarily for its grain, which has numerous uses. It is ground into flour for bread, mixed with wheat to make other flour products, used like rice in soups, roasted as the basis of a coffeelike drink, fermented into an alcoholic beverage, and mixed with honey to make a candy. The young leaves, like those of amaranths, are eaten as a fresh vegetable.

We even cultivate amaranths in the U.S., although most of them are not grown for food. Amaranths are big, showy plants with red and yellow heads and American seed catalogs offer several varieties as flowers. According to Jonathan Sauer, amaranths have been grown in Europe as ornamental plants since at least the eighteenth century.

If amaranths and chenopods furnished the North American Indians with protein-rich grain and nutritious leaves, why can't today's domesticated food species do the same for us? There's some evidence that they can, at least in the case of the amaranths. At John Robson's request, the Rodale Press, which publishes *Organic Gardening*, began growing grain amaranths at its experimental farm in Pennsylvania in 1975, with some 300 of its readers across the country also participating. The results exceeded expectations. In Pennsylvania, the best yield was one thousand pounds of seeds per acre for *A. hypochondriacus*. Elsewhere, however, a reader produced 2.3 tons per acre. The plants grew well (six-foot plants were common), proved hardy even in adverse conditions, and required little attention. Dr. Alex Cunard, Rodale's research director, says one-quarter of an acre could be handled by a single gardener with hand tools. More amaranth seed may be available by the time you read this, but at present only Thompson and Morgan offers the grain amaranth. Burpee, however, sells *A. gangeticus*, also known as Chinese spinach, under the name Tampala. If you want to try a chenopod species, though, you don't even have to grow it; just pick some leaves of the pigweed or lamb's-quarters, a wild species of chenopod found all over the United States, and eat it as a salad green or cooked vegetable.

For the present, the home garden looks like the best place for these nutritious grains, since both have problems that make commercial cultivation unfeasible. Amaranths require hand winnowing and threshing, and quinua meal has a bitter taste due to the presence of substances called saponins. Saponins can be removed with water, and a new saponin-free variety of quinua has been developed in Bolivia, but little information on it is available as yet.

An alternative plant protein source that also looks like a very good candidate for our salads or vegetable side dishes is an aquatic weed called the duckweed that may well be growing wild in a nearby pond. The National Academy of Sciences devotes an enthusiastic appendix to this weed in a booklet called *Making Aquatic Weeds Useful: Some Perspectives for Developing Countries*. Like its other book on alternative sources of protein, *Underexploited Tropical Plants*, this survey emphasizes plants for use in developing countries where protein is scarce, but duckweeds have exciting potential for our own country, too. The duckweed is actually a group of aquatic plants that includes four genera and forty species. Some species are found in fresh waters throughout the world, but the duckweed that serves as human food today is *Wolffia arrhiza*, a minute plant that is eaten as a fresh vegetable in Burma, Laos, and Thailand. Cultivated in rain-fed ponds, it can be harvested every three to four days, during which time it doubles its numbers. It has 20 percent protein on a dry-weight basis.

Wolffia arrhiza hasn't been cultivated in this country as yet, but several other fast-growing duckweeds are being grown successfully in Louisiana by Dr. Dudley D. Culley of Louisiana State University. In outdoor ponds fertilized with farm wastes, two species, *Spirodela oligorhiza* and *Spirodela polyrhiza*, doubled their bulk every three days during the growing season, which lasts about nine months in Louisiana. Culley estimates that if all the duckweeds were harvested, the crude protein produced by one hectare (2.5 acres) of duckweeds would equal that obtained from sixty hectares (150 acres) of soybeans. Another duckweed, *Lemna minor*, doubled its weight in five days in a Louisiana pond. In this last species, the crude protein content is almost 40 percent. Chickens did better on *S. oligorhiza* than they did on their usual diet. The plants grown in the Louisiana experiments are intended for animals, but the fact that people in a few areas of the world eat duckweeds indicates that some of the fast-growing aquatic plants may prove to be a cheap high-protein vegetable for humans in this country.

In addition to plant protein sources on which at least some research attention has already been focused, there are a few that are almost completely unexplored but seem to hold promise in certain situations. Eelgrass (*Zostera marina*) is a grain with 13.2 percent protein, about the same as wheat, and grows in warm, shallow seawater in the Northern Hemisphere. The Seri Indians on the west coast of Mexico once used a flour made from the grain as the basis for a gruel. Arrowhead (*Sagittaria trifolia*), another aquatic plant, produces eight or more edible roots that contain 5 to 7 percent protein, twice as much as well-known root crops. An Asian species, it is regarded as a weed throughout most of its range, but it is used in some Japanese and Chinese dishes. *Acacia albida*, an African tree legume that is highly resistant to drought, bears pods with seeds containing up to 27 percent protein. People in Rhodesia eat them in times of famine. Another nutritious African plant is nara (*Acanthosicyos horrida* Welwitch), a South African cucurbit that grows wild in desert areas and almost rivals the winged bean in its versatility. The pulp is high in protein for a fruit and the lima bean-sized seeds have up to 60 percent protein. According to Dr. J.M.J. de Wet of the University of Illinois, both fruit and seeds are eaten as food, the fruit by various tribes and the seeds by both tribesmen and local farmers. One use of the seeds is to mash them to make a kind of butter.

All of the plant protein species talked about thus far in this chapter could probably become established in the United States only in a limited area. But some other unconventional plants have a better chance of providing a significant amount of our protein in the future, because they are already raised commercially here. The products they furnish today are not protein foods, but each plant is capable of providing ample protein. At least four plants fall into this category: sunflowers, safflowers, guar, and okra. Sunflowers and safflowers look like particularly good protein resources because both have undergone a dramatic expension in acreage here in the last decade. We are the leading producer of safflowers in the world and fourth in world production of sunflowers. Dr. David E. Zimmer, who has an Agricultural Research Service appointment at North Dakota State University in Fargo, a center of sunflower research, says the United States exports more whole sunflower seeds than any other country.

If you thought the tall, handsome sunflower (*Helianthus*

annuus) was just another pretty flower, there's ample excuse for your ignorance. As a food, the sunflower has had its ups and downs in this country. A native of North America, its seed was once eaten by the Indians. But the showy flower, not the seed, caught the Spaniards' fancy back in the sixteenth century, and they took it back to Spain. It quickly became a popular garden flower all over Europe. Meanwhile, back in the land of its origin, the Indians' food source was relegated to a largely ornamental role by the white man, although the seeds retained a small market as a snack. Until the last decade, however, the sunflower's biggest achievement here was being named the state flower of Kansas. In the early twentieth century, a Russian agronomist developed a sunflower with almost twice as much oil as the standard varieties. The emergence of high-oil Russian varieties has led to the establishment of the sunflower as a prime source of oil in the USSR and much of Eastern Europe and to its reestablishment here. The oil has a high proportion of polyunsaturates, making it a desirable food item.

Most of the sunflowers we grow here today are the Russian oil varieties, including one known as Sputnik. We also grow domestic low-oil sunflowers for use as a snack and health food as well as for birdseed. Much of our oil-rich sunflower crop is exported to Europe.

What happens to the residue left over after sunflower oil is extracted is what happens to most of the soybean, cottonseed, and peanut meal after their oils are removed. It's fed to animals, principally ruminants such as the cow. Sunflower meal has about 47 percent protein, a figure that exceeds that of soybean meal. Its PER is 1.64, which is adequate if not impressive, and its BV is 60, above the average for beans. The relatively low PER probably reflects the fact that sunflower meal is low in lysine and isoleucine and contains about 11 percent fiber, a level that rats (and people) find somewhat hard to digest. Another minus for sunflower meal is its color. It's grayish, a shade that is due to the presence of a substance called chlorogenic acid. With these drawbacks, the sunflower may not sound like the best protein source for commercialization, but the big, handsome plant has one major advantage: There are a lot of sunflowers around and there will be more in the future. Acreage planted to sunflowers doubled in the last decade in the United States, with North Dakota and Minnesota the leading producers. Both states are expanding their sunflower plantings, as are a few other states, particularly Texas and California. Some agricultural experts believe sunflowers could become the badly-needed alternative crop to soybeans, particularly in areas where soybeans do not do well because of climate factors.

With the future of sunflowers so sunny in the United States, researchers at a number of institutions are looking at ways to put sunflower protein in the human diet. We already eat whole sunflower seeds, but the seeds used as snacks and in some commercial bakery products come from different varieties than those used for oil and meal. The seed of the nonoil varieties is large, with a hull that can easily be removed. The smaller seed of the oil varieties has a hull that clings so tightly to the seed it cannot be completely removed even in processing. Sunflower seeds sold as birdseed are either small seeds of the non-oilseed varieties or a special birdseed variety. The emphasis in sunflower food research today is not on the edible seeds of the nonoil varieties, but on the meal left over after the oil is extracted from the oil varieties. A group of researchers at the University of Texas received a four-year grant in 1976 to study the use of sunflower flour in human foods. Two foods they expect to test the flour in are bread and snack chips.

At this point, it looks as if sunflower concentrates and isolates may turn out to be a better way to use sunflower protein because they avoid the problems of high fiber and dark color. According to Dr. James A. Robertson of the Richard B. Russell Agricultural Research Center in Athens, Georgia, several processes have already been developed in the laboratory to simultaneously take the chlorogenic acid out of the defatted meal and process the meal into low-fiber concentrates or isolates. The resulting products are light-colored, bland-tasting powders that may make a good protein additive. Feeding trials with rats by a Canadian research group show that one sunflower concentrate has a 2.05 PER compared to 1.64 for the flour, even though the protein content of the two products is only a few percentage points apart. When the concentrate was mixed with peas in the Canadian study, the PER jumped to 3.31, which is higher than that of milk casein.

The safflower (*Carthamus tinctorius*), a native of the Middle East, is another recent agricultural success story here, and for much the same reason as the sunflower. Its polyunsaturated oil is very popular overseas, giving the same impetus to research on its meal that occurs with the sunflower. Safflower protein even has the same drawbacks as sunflower protein: dark color, high fiber, low lysine. With safflower, there's another problem: bitter taste. Nevertheless, safflower has nutritional attractions. After the oil is squeezed out of the seeds in commercial processing, a meal with 45 percent protein and 17 percent fiber remains. Its PER is 1.4, which is nothing to get excited about, but it has a surprisingly high BV of 84.9 to 86.0, based on feeding trials carried out by Dr. George O. Kohler of the U.S. Department of Agriculture's Western Regional Research Laboratory in Berkeley.

Kohler and other researchers at Berkeley have been investigating the safflower for over a decade, seeking ways to turn the protein-rich meal into human and animal foods. One result of their work is a process developed by Kohler and E. E. Goodban for turning the fibrous meal into a white, flour-like substance with 65 percent protein and only 2.2 percent fiber. Another Berkeley researcher, Dr. Antoinette A. Betschart, has developed a safflower protein isolate (SPI) which may turn out to have almost as many desirable functional properties as soy protein isolate. It is readily soluble in liquids, produces a long-lasting foam, holds water adequately, absorbs fat better than soy, and keeps ingredients together in liquids as well as soy does. None of these qualities has anything to do with nutrition, but a food ingredient that boasts all of them is much more likely to be used as an additive or nutritional source in processed foods than one that doesn't.

My own personal vote for the most obscure plant protein source now being grown commercially in this country is guar (*Cyamopsis tetragonoloba*), one of the thirty-six plants described by the National Academy of Sciences in *Underexploited Tropical Plants*. Guar, a legume, has 34 percent protein and 40 percent oil in its seed. Why haven't we heard about a plant with these nutritional attractions? Although guar is grown in Texas and Oklahoma, its present utility here lies in its gum, a thickening agent widely used for both industrial and food purposes. A press cake left over after gum is extracted is fed to cattle. In India, people eat guar pods and seeds. Guar is grown more widely in India than any other country, but Dr. Theodore Hymowitz, the University of Illinois researcher who is working on the winged bean, believes guar was transported there by ship from its original home in the Middle East. Guar seed contains antinutritional factors similar to those in soybeans, but it is believed that they can be eliminated easily by cooking. Research into its nutritional potential is almost completely lacking but with 34 percent protein and 40 percent oil, guar is worth looking into.

Nobody in this country eats guar, but okra (*Hibiscus esculentus*) is a popular green vegetable in the southern states. It's a tasty vegetable, particularly in traditional recipes such as chicken gumbo (gumbo is another word for okra), but when we eat okra as a vegetable, we miss the most nutritious part of the plant. Green okra consists of the immature pods and seeds. When the seeds mature, they have 20.58 percent protein and a balance of essential amino acids closer to that of soybeans than almost any other source of plant protein. According to a study carried out by Drs. Pavlos A. Karakoltsidis and Spiros M. Constantinides of Rhode Island University, okra and soybeans have almost the same high content of lysine, around 8 percent, and the same low content of the sulphur amino acids, cystine and methionine. Levels of all the other essential amino acids were very similar.

This is a significant finding, but okra seed has an even more outstanding feature. When okra seed flour was fed to rats in the Rhode Island study, the animals' weight gain resulted in a PER of 3.4—considerably higher than that of soybeans, which average about 2.3. In fact, a 3.4 PER is higher than that of any other known sources of vegetable protein and about the same as that of milk casein.

The Rhode Island University researchers admit that this figure is difficult to explain but they suggest that "unknown growth factors" may be present in okra seed. It makes no difference, they report, whether the okra seed flour is heated or unheated, a finding that indicates no antinutritional factors are present. In addition, malnourished rats that ate okra seed flour recovered just as fast as malnourished rats that ate casein. It should be noted, though, that another study of okra seed, this one by the same University of Arizona team that studied the buffalo gourd, reported a much lower PER when the seed was fed to mice. The Arizona study also found a lower level of lysine, about 6 percent, as well as different levels of the other amino acids. The discrepancies are partially explained by the fact that the two research teams used different varieties of okra grown in different climates, factors which have been shown to affect amino acid levels in other plants. Only further experimentation, as the Arizona study concludes, will clarify the matter.

9

SCP: PROMISES, PROMISES

> "The little man took the ring, began to whir again at the wheel, and by morning had spun all the straw into gold."
>
> —*Rumpelstiltskin*

"YOU'VE NEVER TRIED SCP? LET'S HAVE SOME."

Dr. Clayton Callihan of Louisiana State University in Baton Rouge, reached down behind his desk and casually pulled out a large plastic bag. He opened it and took out a few yellowish chunks of what looked like pieces of dried natural sponge. "*Cellulomonas*," he said. I took the chunks and nibbled at one cautiously. It was dry, a trifle sticky to chew, and essentially tasteless.

Callihan, who was eating a chunk too, chuckled at my expression. "Now you can say you've eaten bacteria. I guess I've eaten a pound of it altogether."

Cellulomonas is a form of SCP, or single-celled protein, a general term for an alternative source of protein that some scientists consider the best prospect to fill the world protein gap. Besides bacteria, the term SCP covers fungi, yeasts, and algae. All are plant microorganisms with only one cell or, in the case of algae such as the seaweeds or fungi such as the mushrooms, a colony of single cells. In the past decade, food scientists in the United States and abroad have been looking at all these plants with renewed interest. From a nutritional point of view, their attractions are many: fast multiplication, high protein content, good balance of amino acids, low space requirements, and in many cases, growth, on cheap materials such as bagasse (the substance left over after sugar is extracted from sugarcane).

One form of SCP, the algae, will even grow on nothing or what passes for nothing as far as costs are concerned—sunlight and water.

Bacteria come out on top in three of the categories named, giving them a special attraction to nutritionists. Not only are they the fastest-growing microorganisms, on the average (the champion, *Pseudomonas*, doubles its weight in nine minutes), but they have the highest protein content, from 50 to 80 percent, and they grow on more materials than any other form of SCP, including feathers, newsprint, and cow manure. They come in second in amino acid balance, behind some of the algae. You don't like the idea of eating bacteria, which you associate with disease? The overwhelming number of bacteria, like the overwhelming number of all forms of SCP, are harmless. Cheese, sour cream, and some other fermented foods are made with bacteria, and the final product contains large numbers of the tiny organisms. They taste delicious.

Cellulomonas doesn't taste delicious, but it has special attractions of its own. Like some other bacteria, it grows on cellulose, the material that makes up about one-third of the weight of all trees, vines, grasses, and straw. Some 450 million tons of waste cellulose are produced each year in the United States alone. One way to get rid of it is to let cellulose-eating bacteria use it as a substrate, or growth medium, thus converting a waste material into a protein-rich food. *Cellulomonas* has a particular favorite among the many cellulose sources: bagasse. There's a lot of it around, too, some 13 million tons being produced in the United States each year.

"If you haven't seen any bagasse, there's some behind you," said Callihan.

I turned around and saw a bag full of what looked like dry, gray stalks. It is an unimpressive-looking substrate but not, apparently, to *Cellulomonas*, which was first discovered eating bagasse beneath a nearby sugar mill. Most of the bagasse used in the LSU process, according to Callihan, comes from the university's own sugar mill, a scaled-down version used to train agricultural students. To perform the Rumpelstiltskinlike feat of turning this gray mass into protein, the bagasse is chopped up, mixed with water, and put into a fermenter, a large, temperature-controlled metal tank. Enough *Cellulomonas* is added to ensure the viability of the culture, and from then on, it's up to the bacteria or, rather, to their enzymes, which do the actual digesting. In the LSU process, the enzymes turn bagasse into *Cellulomonas* in four to six hours. The soggy mass that comes out of the fermenter is sent to a settling tank, the liquid removed, and the cells dried. For every ten pounds of bagasse that goes into the fermenter, about four pounds of *Cellulomonas* emerges. It has 50 to 55 percent protein, which means that about two pounds of protein is produced for every ten pounds of bagasse.

The fermentation process sounds simple and it is in essence, but, as in most new processes, problems arise. For one thing, *Cellulomonas*, although it digests bagasse, does so very slowly, making the process expensive if Nature is allowed to follow her course. To speed her up, the cellulose in the LSU process undergoes a chemical pretreatment that literally pushes the cellulose molecules apart. "It leaves holes for the bacteria to get in," as Callihan puts it. Another problem is contamination. To make sure no foreign bacteria get into the process, the cellulose must first be sterilized and then the work area where the fermentation takes place kept clean by means of air filters, self-sealing doors, and other precautions.

A number of promising bacterial protein projects, Callihan noted, have been ended prematurely by contamination. "You need very strict controls on this process," he cautioned. "It's not that the bacteria would hurt anyone, but that they can be attacked by other bacteria."

Occasionally, the presence of foreign bacteria actually helps the digestive process along. A few years ago, the LSU team noted that on several run-throughs of their project, the cellulose was being devoured at a much faster rate than usual. When the contents of the fermenter were examined, a bacterium called *Alcaligenes faecalis* was discovered. The intruder, it's believed, eats a product that inhibits *Cellulomonas*'s enzyme activity, allowing digestion to proceed without interruption. *Alcaligenes faecalis* has about 78 percent protein compared to no more than 55 percent for *Cellulomonas,* and a slightly better balance of amino acids. The LSU researchers now use mixed cultures of the two bacteria in some of their work.

How does a bacterial protein like *Cellulomonas* stack up against our number one alternative protein source, the soybean, as far as nutrition goes?

With 50 to 55 percent protein, *Cellulomonas* obviously has a protein level superior to that of soybean meal and flour, which are about 43 percent protein. The bacteria's amino acid balance isn't quite as good as soy's, being a little lower in the essential amino acids tyrosine and cystine, but it's better in one respect: methionine. *Cellulomonas,* with 1.8 percent methionine, comes closer to the 2.2 percent reference standard set by the FAO than soy with 1.1 to 1.8 percent. As far as digestibility goes, *Cellulomonas* lags behind soy. One of the few feeding trials conducted with the yellowish bacteria shows that rats on a diet with 10 percent *Cellulomonas* protein gained much less weight than rats eating a diet with 10 percent milk casein protein. Interestingly enough, though, rats that ate a diet with 40 percent *Cellulomonas* gained more than those that ate the 10 percent diet (although still not as much as the rats that ate the 10 percent casein diet).

In spite of the nutritional shortcomings of *Cellulomonas,* the LSU process is considered one of the most sophisticated and successful efforts to produce microbial protein in this country. Scientific journals refer to it regularly and Callihan and other LSU faculty members involved in the project are invited to give papers at scientific conferences. When I visited LSU, Callihan had just finished a paper he was to give the following month in Pakistan. In the light of all this approbation, it's a little disconcerting to report that *Cellulomonas* has, in effect, been sent back to the sugar mills. A large pilot plant the university built with NASA cooperation has been partially dismantled, and a widely publicized plan by the Bechtel Corporation of Los Angeles to put the system into commercial production is no longer under consideration. When I talked with Callihan, he didn't have a single graduate student working on the project.

What went wrong with *Cellulomonas*?

Callihan, a smiling, joke-cracking man, stops smiling when he thinks about the present status of the LSU bacterial protein project. He ascribes the stagnation to a combination of university politics, a poorly controlled pilot plant put into operation by Bechtel on the West Coast without Callihan's assistance, and the results of a nutritional study by a group of California researchers. The nutritional study may be the most serious problem. It was carried out by Dr. Carol I. Waslien, Dr. Doris H. Calloway, and Dr. Sheldon Margen, M.D., all of the Department of Nutritional Sciences at the University of California. The results indicate that *Hydrogenomonas eutropha*, a bacterium with 70 percent protein, and *Aerobacter aerogenes*, another bacterium, produce a fairly wide range of minor but unpleasant symptoms in some human subjects. Among them are diarrhea, nausea, headache, weakness, rashes, and pains in the arms and legs. Animals that ate the same bacterial proteins had no symptoms, even when they ate far more than the human subjects. *Cellulomonas* was not tested, but the study aroused the fear that it might bring on the same effects in humans.

Callihan refers to the California research by a one-syllable word that is used as a substrate for bacteria by another researcher, but the study is taken more seriously by some other scientists. The probable culprit, they believe, is a substance called nucleic acids, which have been shown to cause elevations of uric acids in humans if used over an extended period of time. Raised uric acid levels lead to gout, kidney stones, and gallstones. Some bacteria contain from 15 to 16 percent nucleic acids, a fairly high level. Yeast and fungi contain from 6 to 11 percent, still a high level. Algae have less. A number of new processes now exist to remove nucleic acids or decrease their level in SCP, but thus far none has been used outside the laboratory. For the present, caution nutritionists, it's safer to restrict consumption of protein sources containing high levels of nucleic acids. The suggested level for nucleic acids is 2 grams (.07 ounce) per day, which effectively restricts bacteria, fungi, and yeast to the role of additives for the present. Further nutritional studies with bacterial protein will undoubtedly be needed before larger amounts of SCP can be consumed.

Bacterial animal feeds, however, do not face the same nutritional restrictions as bacterial human foods. Animals metabolize purine, the chemical that is the precursor of uric acid in humans, in a different way than we do, allowing them to eat a high level of nucleic acids without ill effects. Dr. W. Dexter Bellamy, a biochemist with General Electric Company's Research and Development Center in Schenectady, New York, has been working on a bacterial animal food for a number of years. He uses a substrate even cheaper than bagasse: cow manure. Manure contains a large component of cellulose, making it attractive to certain heat-loving bacteria known as thermophiles. Bellamy found some of them in Yellowstone Park, near the hot springs. When the thermophiles were put in a fermenter to which chemically treated manure from a feedlot used to fatten beef cattle had been added, the little organisms produced a substance with high levels of lysine and the sulphur amino acids methionine and cystine. Most forms of SCP are low in the sulphur amino acids.

The process looked so promising in Bellamy's laboratory that General Electric built a large pilot plant in Casa Grande, Arizona, which it opened with a blizzard of press releases and a ceremony attended by local dignitaries. The plant was designed to process the manure of one hundred cows a day and turn it into a high-protein supplement for animals similar to soybean meal. But before a single chicken or pig had eaten the bacterial food, the plant closed due to both economic and technical factors (including contamination of cultures). Meanwhile, back in the laboratory, Bellamy and his colleagues continue to work on bacterial protein. Will the Casa Grande plant ever reopen? "It depends," Bellamy told me, "on the price of soybeans as much as any other factor."

The present status of these two highly regarded projects indicates the large gulf that exists between the promise of SCP and the reality of putting it on the American dinner table, or even in the American feed trough. Scientists are genuinely excited about SCP as a means of supplying protein to a world with an increasing population and a shrinking number of acres on which to raise food by means of conventional agriculture. "It has been calculated that a 10 percent supplement to the world's food supply could be provided by an SCP fermenter occupying an area equivalent to about 1.5 square kilometers of the earth's surface," said a 1975 National Science Foundation survey of protein resources. In theory, high-quality SCP can be produced fast in minimal space on cheap substrates, some of which might otherwise become a disposal problem. But if SCP is so good, why aren't we pushing ahead with a program to put it in our diet, or at least in our livestock's diet, as fast as possible?

One reason, doubts about nutritional value, has already been mentioned in connection with *Cellulomonas*. The same doubts apply to other forms of SCP, although in somewhat lesser degree. But SCP has other problems to overcome before it can become a significant protein resource here. According to microbiologist Dr. John H. Litchfield of Battelle Memorial Institute in Columbus, Ohio, the success or failure of any unconventional source of protein depends on its ability to compete with conventional sources on three counts: nutrition, acceptability (by people and animals), and cost. In this country, he points out, no source of SCP comes out ahead on all three counts compared with a relatively cheap source of high-quality feed protein such as soybeans or cottonseed meal. In 1974, Litchfield published a study that shows that the least expensive forms of SCP still cost a little more than soybean or cotton-seed meal on a price-per-kilogram basis and fall a little short of both on nutritional value. With those two handicaps, SCP finds doors closed to it even in the animal feed industry. In 1977, we didn't have a single commercial facility producing SCP protein for animals.

Incidentally, one of the cheapest forms of SCP Litchfield found was *Cellulomonas*. The others were yeast grown on a paper-industry waste and bacteria grown on methanol, a cheap fluid derived from methane.

In Europe, the SCP picture is completely different. A survey of commercial microbial protein operations presented at the 1975 annual meeting of the American Chemical Society by Jeremy Wells of the McKee CTIP Corporation shows eight plants in operation throughout the world, seven of them in Europe. The biggest, which produces from three hundred thousand to five hundred thousand tons of yeast annually on a derivative of hydrocarbons, is in the Soviet Union. Only one plant, Amoco Food Company's Hutchinson, Minnesota plant, is in the United States. It produces yeast on ethanol, which is ethyl alcohol, a food-grade derivative of hydrocarbons. Unlike all the European plants now in operation, Amoco sells its yeast to food processors for human use. The European plants sell their products as animal feed. At the time Wells did his study, three more plants were slated to go into operation in Europe shortly. With one exception, the eight plants and the three about to open grow yeast, most of them on gas-oil, a semipurified derivative of hydrocarbons, or the more expensive "n-paraffins," a refined derivative of hydrocarbons. The exception is a plant in Finland that grows fungi on sulfite wastes.

Europe is embracing SCP with so much more enthusiasm than we are because it lacks our vegetable protein resources. With a 50 percent increase in meat production over the past decade, Europe needs vast quantities of protein supplements to augment feed rations. European nations buy some soybeans from the United States and other soybean-producing countries, but imported soybeans cost European farmers more than home-grown soybeans cost our farmers. The higher soybean prices in Europe make SCP more competitive there than it is here. Actually, as Wells noted in his paper, most European farmers in 1975 were not feeding SCP as a soybean replacement but as a milk replacement in calf rations. But microbial proteins are expected to displace some of the soybeans before long and commercial operations are designed and built on such expectations.

John Litchfield believes Europe offers a preview of what we may be doing here with SCP in the future. As the price of soybeans rises, he says, SCP will become more competitive as an animal feed ingredient, particularly those SCP forms grown on renewable substrates such as cellulose wastes, sugars, and starches. Meanwhile, research will produce improved versons of today's microbial proteins with qualities such as faster growth and lower levels of nucleic acid. The development of isolates and concentrates will offer an additional range of products for both supplementation and functional uses. Already, Litchfield points out, laboratory processes exist to accomplish all of the above. Nor does he rule out the use of SCP in human diets. On a laboratory scale, he has produced fibers spun from yeast concentrates and algae concentrates that he believes would make a suitable meat analog. Another group of U.S. researchers has performed the same feat with a bacterial concentrate. The bacterial fibers were stronger than spun soy fibers. One big advantage of SCP concentrates is that they can be processed to completely remove nucleic acids.

If we're going to be eating microbial protein in the future, it might be a good idea to take a closer look at these microorganisms, which are as remote to most of us as the features of a distant planet. Bacteria are the smallest of the four forms of SCP. Their cell size is so minute that you could put a trillion in a tablespoon. To study them requires an electron microscope, an expensive instrument that can enlarge organisms far beyond the capacities of the light microscope. There are three principal types of bacteria, the division being made according to shape: rods, spheres, and spirals. Reproduction is princi-

176 /   FUTURE FOOD

pally by division, although some bacteria form spores that
undergo a form of sexual reproduction. As mentioned earlier,
bacteria grow very rapidly. *Pseudomonas* is the leader, multiplying itself every nine minutes, but many bacteria replicate
themselves in twenty minutes or so. The protein content of
bacteria ranges from about 50 to 70 percent and bacteria have
a better balance of amino acids than do yeasts or other fungi.
Bacteria will grow on almost anything, including a variety of
cheap cellulose wastes, although their growth on cellulose
tends to be slower than on many substrates.

BIOLOGICAL VALUES OF SOME SINGLE-CELLED ORGANISMS

ORGANISM	BV
Algae	
Chorella ellipsoidea and *Scenedesmus*	54.3
Spirulina maxima	72
Bacteria	
Micrococcus cerificans	76
Hydrogenomonas eutropha	77.6±4.6
Fungi	
Fusarium species	70–75
Yeast	
Candida lipolytica	
Alkanes	61
Gas-oil	54
Candida utilis	32–48

Source: E. S. Lipinsky and J. H. Litchfield, 1974.

Their drawbacks are small size and low densities, which
make harvesting difficult; the highest nucleic acid level among
the SCP's; and a poor image. Most of us associate bacteria
with disease. Some certainly do cause dreaded diseases—tuber-

culosis and cholera, among others—but the vast majority of bacteria are harmless. Some are helpful to man. Millions of bacteria live on the human skin, for instance, where they help keep that surface clean by devouring less friendly organisms. Other bacteria are used in making fermented foods, including cheese, buttermilk, and yogurt. A list of foods that incorporate healthy amounts of bacteria includes those foods, as well as butter, rye bread, sauerkraut, pickles, vinegar, and fermented sausage.

And while we're talking about bacteria in foods, let's not forget silage, the cow's cold weather staple, which is simply hay fermented in a silo by means of bacteria.

The fungi and yeasts are often discussed separately as if they were distinct forms of SCP, but yeasts are a special kind of fungi. Another word for fungi is mold. Unlike the minute bacteria, some fungi are readily visible to the naked eye. Toadstools and mushrooms are fungi as is the truffle, a delicacy very popular in France. How can these fairly large plants be single-celled organisms? They are actually colonies of single-celled organisms rather than a single multicelled organism. At another stage of their life, some of these colonies are dispersed in separate cells. Many fungi reproduce by sending out spores that travel on the wind or in the water, as some fungi are aquatic. Sexual reproduction takes place between the spores in some species. The yeasts have their own specialized form of reproduction known as budding. They produce a group of little protuberances that eventually break away to form new individuals. Some fungi other than yeasts have 12 to 50 percent protein, with mushrooms at the upper end of the scale. Yeasts have about 60 percent protein. Nutrition figures on fungi are sparse, but one species has a BV of 70 to 75 percent, a fairly impressive figure. Yeasts, rather surprisingly, make a poorer showing than other fungi on the whole, the BV's in one group of *Candida lipolytica*, a popular food and feed yeast, ranging from 31 to 61. According to another much-used nutritional measurement, PER, which is based on rat feeding tests, a strain of *Candida utilis*, which is probably the most popular food yeast, ranged from 0.9 to 1.4. These are low figures but supplementation with methionine, the amino acid in which yeast is low, raises them dramatically. Supplemented *C. lipolytica* has BV's of 91 and 96 on different hydrocarbon substrates and *C. utilis*, 88. Some processes yield better-quality yeasts. Amoco claims its food-grade *Candida utilis* grown on ethanol has a PER of 1.7 without supplementation.

Both fungi and yeasts have some advantages over bacteria as food. Yeasts have about 6 to 11 percent nucleic acids, fungi even less. Both grow more slowly than bacteria but their cells are larger and yeasts mass more densely in cultures, making them easier to harvest. Also, people know and like them. Mushrooms are popular all over the world, truffles bring high prices from gourmets, and yeasts are a standard in the Western kitchen for bread baking. Yeasts are used today in small amounts in some foods as a nutritional or functional additive and health food stores sell an add-it-yourself variety for additional supplementation. In times of crisis, yeasts have been widely used as foods. During World War II, Germany grew yeast on sulfite liquor and wood sugar for food use. According to microbiologist Bernard Dixon, author of *Magnificent Microbes* and editor of a British science journal, at least eight yeast plants were operating at the height of the war, producing sixteen million kilograms of *C. utilis*. All of it was incorporated into human food.

Aside from times of crisis, though, neither fungi nor yeasts have formed a significant part of the human diet. What will be the nutritional result if we eat large amounts of fungi? The findings from human and animal tests are somewhat confused. In one series of tests, animals showed decreased feed consumption and increased excretion of feces when more than 15 percent of the diet was made up of yeasts. In another yeast feeding test, this one with humans, the same University of California group that carried out the bacteria nutritional study mentioned earlier found "dangerously elevated" blood and uric acid levels in the subjects' urine when yeasts were a major source of protein. On the other hand, a few animal tests with fungi show results as good as those with meat, even when all the protein in the diet came from fungi. Obviously, only further research will clarify the nutritional values of fungi and yeasts, and pinpoint which, if any, can be eaten in significant amounts without prior removal of nucleic acids.

Many fungi are sizable plants, but some algae are very large indeed. The largest plant ever recorded, a seaweed called *Macrocystis*, was seven hundred feet long. This huge plant is not a single cell but a colony of single cells. *Macrocystis*, however, belongs to a family of seaweeds that includes a species with the largest cells of any plant or any animal, for that matter. Each cell of this seaweed is several feet long and comprises a single, one-celled plant. At the other end of the size scale are the tiny blue-green algae. All algae multiply by division, grow rather slowly for microorganisms (but fast for plants), are exclusively aquatic, and like sun and warmth. From a food standpoint, they possess undoubted attractions, so much so that some scientists think they are *the* SCP of the future. One advantage they possess is that they need no substrate, just water, sunlight, and warmth. Some grow well on diluted sewage.

Other points in algae's favor are large cell size, low nucleic acid content and, in the case of some species, very high nutritive value. *Spirulina maxima*, a blue-green algae, has 65 to 70 percent protein, a 2.3 to 2.6 PER, the highest among the microbial protein sources, and a 72 BV. Its nucleic acid content is 4.2, low for an SCP source.

Drawbacks? The major one is the fact that algae can only be grown where there is plenty of sun, water, and warmth. For practical purposes, this restricts them to areas between latitudes 35 degrees north and 35 degrees south. Other black marks are the tough cell walls, which make it hard for humans to digest them, a bitter taste, and a green color. Not all species have the first two qualities, but all are green. Greenness may not sound like much of a drawback, but when you dry algae for use as a food supplement, the product for which it is usually suggested, you get a green powder, which creates the same sort of problems in traditional foods that green leaf protein does. Also, algae brought on digestive upsets in humans in one feeding test and caused increased uric acid excretion in another.

None of this bothers the people on the shore of Lake Chad in Africa, who have eaten *Spirulina maxima* for as long as anyone remembers. The organisms are collected from the surface of the lake with a large vessel and poured into a cloth bag. After the water has drained out, the residue is dried in the sun and then cut into blocks. It's eaten as a green vegetable. Seaweed, a form of algae, is eaten in Japan, and some Pacific islanders eat several species of algae. Powdered algae and seaweed are both sold in health food stores, giving Westerners some exposure to these microorganisms.

Which of the four forms of SCP is likely to turn up first on the American dinner table? Judging from the commercial SCP processes now in existence, the answer is yeasts. Seven out of the eight commercial SCP processes make yeasts and the three slated to begin operation shortly also will make yeasts. Amoco Food Company, one of the seven, is the only United States producer of commercial SCP, and, in 1977, the only one making it for human use. Since the Amoco operation may be a prototype of the kind of SCP plant that will produce a significant part of our protein in the future, a quick look at it is worthwhile.

Amoco grows *Candida utilis,* a form of torula yeast and one of the four yeasts used in food. The others are baker's, brewer's, and fragilis. Baker's yeast is used in bakery products, brewer's (a by-product of the beer industry) in health foods and additives, fragilis in health foods and additives, torula in additives and as a flavor enhancer. Traditionally, torula has been grown on sulfite liquor, or on sugar sources such as molasses. The new wrinkle which Amoco adds to this old process is the substrate. Amoco grows its torula, to which it gives the brand name Torutein, on ethanol, or alcohol. Ethanol can be produced either by fermentation of carbohydrates or by extraction from ethylene, a hydrocarbon product. Since Amoco is a subsidiary of Standard Oil of New Jersey, it gets its ethanol via the latter method.

Why market another yeast when there are a number of food yeasts already around? Amoco believes Torutein offers advantages over other yeasts because it adds less yeasty, bitter flavor to foods. Some food processing firms apparently agree because about one hundred of them were using Torutein in 1977, two years after the Amoco plant opened. These firms use the product principally as a functional ingredient. Yeasts do nice things for food like keeping oil and water together in a mixture, enhancing flavors, thickening gravies, adding opaqueness, and increasing volume. Amoco now offers four food yeasts: the standard yeast, a yeast with even less flavor, and two yeasts with varying functional qualities for special uses. Standard Torutein has a PER of 1.7 and 9 percent nucleic acids. A low-nucleic acid yeast, according to Amoco, is now being developed.

Yeasts obviously offer many advantages, but the hydro-carbon-derived substrate which Amoco and most other commercial SCP manufacturers use faces a big problem today. The substrates are increasingly expensive and scarce. When research on hydrocarbon substrates for yeast was begun some years ago by various oil companies, hydrocarbons looked like a fine substrate, being cheap, available, and transportable. Now the future holds nothing but steadily rising prices and dwindling supplies. The best thing about this situation from the standpoint of the world protein shortage is that hydro-carbon-grown yeast looks like a promising way to add protein to the diet of people in Middle Eastern countries, which have most of the world's hydrocarbons but few of its protein resources. As yet, no SCP plants are under way in the Middle East, but a British oil firm has announced plans for one in Venezuela, one of the world's leading oil producers.

What will we grow our SCP on in oil-short countries of the Western world? John Litchfield believes that methane gas, which has been going down in price as hydrocarbons go up, and methanol, a liquid into which methane can be converted, are the most attractive nonrenewable substrates. Some bacteria flourish on methane and methanol, a proclivity that may help push them ahead of yeasts as an SCP source. Not only that, but the bacteria also grow well on some renewable substrates, which Litchfield and some other experts see as more attractive for the future. Renewable substrates include sunlight and water, waste cellulose, sugars, and starches. No commercial process yet exists to grow bacteria on either renewable or nonrenewable substrates, but there are a number of very promising research projects with cellulose-eating bacteria, led by the LSU and General Electric projects.

One of these projects is almost ready to begin commercial operations. ICI, a British firm, has developed an efficient process for cultivating *Pseudomonas*, the world's fastest-growing bacteria, on methanol. According to Bernard Dixon, the organism was discovered living beneath an industrial methanol plant in Britain. Transferred to the laboratory, it continued its methanol-eating ways with enthusiasm and later was selected for the star role in a one thousand-metric-ton pilot plant in Britain. The final product is so nutritious that when used in one test as 10 percent of a pig diet, the pigs actually did better on *Pseudomonas* than on their regular fare. A plant with a capacity of one million kilograms of protein per year has been commissioned by ICI for the late 1970's. For the near future, it will turn out only animal feed.

A whole spectrum of bacteria is being used to produce protein on renewable substrates, including newsprint, meat-packing wastes, wheat bran, a watery waste left over after cheese making, a weed that flourishes in the American Southwest, and a mixture of hydrogen, carbon dioxide, and urea that would be available in spacecraft. Bacteria turn all of them into a protein-rich mass, although yields are not always high enough to justify a commercial process. One of the substrates that does result in high yields is mesquite, the dense, thorny plant that infests many areas of the southwestern United States. Prof. D. W. Thayer of Texas Tech University in Lubbock, found that strains of *Pseudomonas* produce a dense, 64.9 percent protein mass from mesquite that exceeds or equals the FAO reference protein in eight essential amino acids. Thayer's principal interest, however, is not in protein but in a complete cattle feed that has a high calorie content. The feed that results from his process is a mixture of bacterial protein and partially digested mesquite containing about 12.9 percent protein.

Fungi and yeasts aren't quite as versatile as bacteria, but some grow well on an odd assortment of substrates, renewable and nonrenewable. One yeast of the *Candida* group grows on acid chemical wastes, producing a mass of cells that is 67 percent protein. Another *Candida* yeast flourishes on waste polyethylene, still another turns tapioca, the delicious but low-protein carbohydrate dessert, into a 55 percent protein entree, while a fourth member of this hard-working family performs the fairy tale-like task of turning straw into protein. Fungi also produce protein from straw, as well as from coffee waste water, pea and corn wastes, cereal processing wastes, and carob bean extract (the carob bean is widely grown in the eastern Mediterranean region). A fungal process that transmutes sulfite wastes from the paper industry into protein has already been commercialized, as mentioned earlier. All these substrates are renewable or else are wastes, ensuring fungi a place on the food shelf even if hydrocarbon substrates become too expensive.

An ingenious fungal process for converting straw into oyster mushrooms was described at the 1976 annual meeting of the American Chemical Society in Berkeley by Dr. Ralph Kurtzman of the USDA's Western Regional Research Laboratory in Berkeley. As a substrate, he used rice straw, the part of the plant left over after the grain is removed. It is usually burned in the fields. The oyster mushrooms—*Pleurotos ostreatus*—produce one ton of fresh mushrooms from one ton of dry straw. As a bonus, the process leaves behind a straw suitable for cattle feed because most of the lignin, a substance which not even cows can digest, has been digested by the fungi. Oyster mushrooms, which contain about 30 percent protein

YEAST PROTEIN PROFILE
Grams essential amino acids per 100 grams true protein.

	AMOCO TORULA YEAST	FAO * STANDARD
Isoleucine	5.0	4.2
Leucine	8.0	4.8
Lysine	7.8	4.2
Methionine	1.3	2.2
Cystine	0.8	2.0
Phenylalanine	4.6	2.8
Threonine	5.6	2.8
Tryptophan	1.4	1.4
Valine	6.0	4.2

* Food and Agricultural Organization of the United Nations.

Source: Amoco Foods Company, Chicago, Illinois.

dry weight, are a pleasant-tasting mushroom grown commercially for table use in Europe and Asia. If all the rice straw in the United States were used to produce oyster mushrooms, Kurtzman told his audience, about two hundred pounds of mushrooms could be produced for every person in the United States.

"Even the most ardent mushroom lover might find that somewhat more than he cared to eat," added Kurtzman. "However, it does illustrate how much energy is wasted. Since other agricultural wastes and waste paper could also be used, the potential for more food becomes staggering."

When it comes to economical substrates, however, the algae are in a class by themselves. All they ask for is sun-warmed water. Even high salinity or alkalinity doesn't bother them, although seawater has too much magnesium for some algae, particularly the nutritious *Spirulina maxima*. True, algae flourish only where it's warm, but then so do oranges and bananas and we find those warmth-loving fruits in our supermarkets. No commercial algae process is in existence, but some of the research projects involving algae look too good to pass up. According to the 1975 National Academy of Sciences publication *Underexploited Tropical Plants with Promising Economic Value*, a pilot plant at Texcoco, Mexico, built around large artificial basins of water produces about one ton of *Spirulina* daily, which is already being sold as a chicken feed additive. The high protein content of up to 72 percent and the 2.3 PER make it the most nutritious of the SCP sources, especially considering its low nucleic acid content of 4.1 percent. *Spirulina* is, of course, green, but the French researchers who have done most of the work on the algae at the Mexican site claim that up to 10 percent of it can be added to cereals and other food products without changing the flavor. That's a big advantage in Mexico, where the diet of many people is based largely on a low-protein cereal, corn.

As yet, however, the *Spirulina* produced by the pilot plant is not being eaten by people, even though history tells us the Spaniards found the Aztecs in Mexico eating one of the *Spirulina* species in the sixteenth century. Before the algae are considered safe for people today, cautions the NAS, "multi-generation feeding studies with animals and prolonged feeding tests with humans" are required.

As far as we know, *Spirulina* is the world's most nutritious algae, but some other species also look enticing. In his 1975 book *Hothouse Earth*, Dr. Howard A. Wilcox, director of the U.S. Navy's Ocean Farm Project, describes an open-sea farm in which the crop will be kelp, the large seaweed. Wilcox paints a glowing picture of a future in which these ocean farms will produce vast amounts of kelp that will be turned into human foods ranging from artificial steaks to cookie mixes. An experimental farm with one species of kelp, *Macrocystis pyrifera* (a relative of the seven-hundred-foot-long seaweed that is the world's largest recorded plant), was set up in 1974 off California's San Clemente Island, some sixty miles west of San Diego. Given a mesh network to grow on, the kelp flourished, growing to an average length of one hundred feet. As often happens with experimental projects, though, something went wrong—in this case, the unauthorized passage of a large ship through the area, destroying the mesh and the farm. Undaunted by this disaster, Wilcox envisions kelp farms covering from 30 to 70 percent of the ocean's surface by the year 2100 and supporting even a population of forty to fifty billion people with the greater part of their daily food and energy needs.

Now there's a vision!

Algae experts tend to be optimistic. Biologist Victor Kollman of the University of California's Los Alamos Scientific Laboratory (LASL) says we can feed thousands of times more people than the world's present population of five billion by using just 1 percent of the world's land to grow algae. The key to Kollman's prediction is a speedup of the normal photosynthesis process produced by aiming a laser at the algae's chlorophyll system. Repeated experiments at LASL indicate that it is, indeed, possible to grow algae faster than nature does, at least in the laboratory. Kollman sees a typical algae farm of the future as a watery plot of about an acre, excavated to a depth of one and one-half feet, and covered with plastic. Solar heat or some other form of heat would maintain a temperature of about 100 degrees Fahrenheit. The advantage to Kollman's plan is that it could be used almost anywhere in the world, not just below 35 degrees latitude.

Kollman does see a problem in public acceptance of all that algae, but he points out that other foods that once were strange are now accepted. Another University of California scientist, however, may have a better way to make algae more pleasing to the taste buds of the jaded Westerner: artificial flavors. "Algae is cheap, abundant and nutritious," says Dr. Walter Jennings, a food flavor chemist at the Davis campus, "but the problem is, it tastes terrible." He believes that adding a flavor such as steak to a textured form of algae would make people relish eating it. At present, Jennings and his assistants are working to isolate and identify chemical compounds that give individual foods their unique flavors. Such compounds, when found, can be reproduced in the laboratory to give algae the exact taste of foods such as steak or lasagne, another Jennings favorite.

Could it be that algae enthusiasts tend to go overboard concerning their watery protein source? Possibly. But it's easy to get worked up about SCP. In all its forms, it holds enormous promise for filling the world protein gap.

10

BEEF LOVE

Address your prayers to the excellent bull, to the principle of all good, the source of all abundance.

—ANCIENT PERSIAN HYMN

WHEN YOU THINK ABOUT IT, THE COW YOU SEE PLACIDLY grazing on a hillside is a remarkable beast. The term "sacred cow" has real meaning. A number of ancient civilizations regarded the cow as sacred and it is still the subject of veneration among Hindus in India. Even in the Western world, various customs and beliefs center around the cow. According to Anthony S. Mercatante's *Zoo of the Gods*, the Spanish bullfight epitomizes the bull as a symbol of brutality and darkness. In the United States, we have a more benevolent cow symbol—Elsie, the Borden trademark. To Mercatante, Elsie reflects all the life- and health-giving qualities of the goddesses Hathor, Isis, and Demeter, each of whom was associated with the cow. But the cow's services are by no means confined to the spiritual. It furnishes meat, milk, hides, fertilizer, gelatin, and even medical products, such as insulin. Man could have survived without the cow—the Eskimos have no cows and neither do some tribes in Africa—but the world would have been a bleaker place without it.

Here in the materialistic United States, we've had a long love affair with the cow, but for the most part, it's not the cow's

PER CAPITA BEEF CONSUMPTION
(pounds)

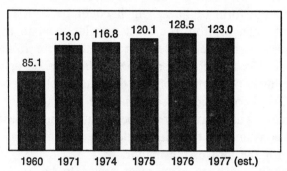

Source: American Meat Institute, 1977.

spiritual aspects that appeal to us but its fleshly ones. In 1976, according to the American Meat Institute, we produced 39.5 billion pounds of meat (excluding poultry), which works out to the highest per capita figure of all time: 192 pounds. Beef led the list with 128.5 pounds per capita, a new record. Pork is well behind with 57.8 pounds per capita. Our meat-eating habit, which is really our beef-eating habit, puts us way ahead of most countries in the world, but not all of them. In 1974, again according to the American Meat Institute, the per capita consumption of meat was 202 pounds in Argentina, 206 pounds in New Zealand, and 209 pounds in Australia, all of them big beef-eating countries, too. Those three countries are also among the world's principal exporters of meat, along with Denmark and the Netherlands. The United States imports far more meat than it exports, the figures for 1975 being 2,237,500 pounds for imports versus 268,400 pounds for exports. As you might guess, most of those imports were beef and veal (young beef): 1,781,800 pounds.

Beef wasn't always our favorite meat. To celebrate the Bicentennial in 1976, the Institute of Food Technologists, a professional society whose members work in the processed food field, put out a booklet called "Food of Our Fathers," which details the food consumption habits in the America of 1776. In those days, the most popular domesticated meat was pork, usually in the form of ham, which could be kept for long periods of time. Mutton, the most popular meat in the Old World, also was widely eaten here. Cattle were raised but largely for milk and cheese, not meat. Game made up a surprisingly large proportion of the diet and included rabbits, squirrels, wild fowl and, as settlers moved west, buffalo. The abundance of game gave our forebears much more meat than they had consumed in the Old World, where game was largely reserved for the lord of the manor.

This early dependence on meat seems to have fixed the meat-eating habit firmly in our minds and stomachs, but our taste for the cow developed later, as a result of the abundance of rangeland available in the newly opened West. In the 1950's, beef ambled into first place among the meats in this country; it's been there ever since. The per capita beef consumption figure doubled from 1920 to 1975, while that of lamb shrank to less than half of what it was in 1920. Pork dropped from 63.5 pounds per capita to 54.7 pounds per capita in the same period. Chicken consumption, like beef consumption, has climbed. In 1920, we ate a mere 13.7 pounds of chicken per capita; in 1973, it was over 40 pounds. When I was a child, the chicken was something of a luxury item, like roast beef, to be enjoyed only on Sundays and other special days. Today, thanks to chicken management practices in which the fowl are raised under tight space restrictions and on rigidly controlled diets, chicken has become one of the most economical meats. The American consumer, noting its price, has made chicken an everyday meat. The turkey, too, has won a more secure place in our stomachs. In 1930, the first year for which figures are available, we ate only 1.5 pounds of turkey per capita; in 1975, it was 8.6 pounds.

Americans have been criticized of late for their level of meat consumption, but there's no doubt that meat is an excellent protein source. Cuts of raw meat have from 14.8 to 20.7

percent protein, cuts of cooked meat, which has lost much of its fat, from 19.7 percent, for a broiled porterhouse steak, to 28.6 percent, for a broiled round steak. Raw bacon has 8.4 percent protein, cooked bacon 30.4 percent protein. If you consider simply the cooked lean portion of the meat, the statistics are even more impressive. The lean portion of a pork loin has 34.6 percent protein, a lean loin veal chop 34.2 percent protein, and a lean veal round cutlet 38.1 percent protein. A braised beef kidney has 33 percent protein. In general, organ meats—kidneys, hearts, livers, and so forth—are higher in protein than flesh meats. Among the meats and meat products comparatively low in protein are frankfurters with 12.4 percent protein and beef brains with 11.5 percent protein. One serving of lean meat weighing about three and one-half ounces will supply about half the recommended daily allowance of protein.

But percent of protein isn't the whole story in foods. For a single protein source to be utilized by the human body, it has to contain all eight of the essential amino acids in certain

ESSENTIAL AMINO ACID
CONTENT * OF POPULAR MEATS

AMINO ACID	BEEF	PORK	LAMB	PROCESSED
Leucine	8.4	7.5	7.4	7.4
Valine	5.7	5.0	5.0	5.2
Isoleucine	5.1	4.9	4.8	4.9
Methionine	2.3	2.5	2.3	2.2
Threonine	4.0	5.1	4.9	3.9
Phenylalanine	4.0	4.1	3.9	4.0
Lysine	8.4	7.8	7.7	7.4
Tryptophan	1.1	1.4	1.3	1.0
Histidine	2.9	3.2	2.7	2.8

* As % of protein.

Source: Kiernat *et al.*, 1964.

amounts. All forms of meat, as well as poultry, wild game, and fish, possess virtually the same lineup of amino acids, including all the essential ones. Meat is somewhat low in methionine, but slightly more than an ounce of beef supplies the 1.10 grams of methionine needed by young men, according to the classic study carried out some years ago by W. C. Rose. If you eat 200 grams—about 7 ounces—of meat, you will get all the essential amino acids you need. So close is meat to the optimum balance of essential amino acids that it is considered a "complete" protein, along with eggs and milk.

Meat also is highly digestible. One way in which digestibility is expressed is the protein efficiency ratio (PER). The egg has the highest PER of any food, 3.8, but meat is not too far behind. Pork tenderloin has a PER of 3.3, beef muscle 3.2, and beef heart 3.1. Another way of expressing the digestibility of meat is the biological value (BV), which depends basically on amino acid balance. The egg has the highest BV, followed by milk, corn germ, liver, and beef.

All meats have much the same amino acid balance but beef obviously has a value to Americans beyond that of nutrition. Have you ever seen anyone drooling over a tasty piece of fried chicken the way people do over a rare steak? Does roast pork, be it ever so delicious, have the same appeal as roast beef? Not in most American households. Given this bias, it's no wonder the livestock industry and the American consumer both react as if someone had killed a sacred cow when it is suggested, as it has been recently, that we phase out the cow and feed the grain that now goes into its capacious four-stomach system to people. Substitute chicken for beef? Beans for steak? Never! On the face of it, though, the argument for getting rid of the cow has a good deal of logic. Cows are hefty animals—at slaughter, the average cow weighs about one thousand pounds —and they eat a lot of food. Wayne Anderson, managing editor of the feed industry magazine *Feedstuffs*, says beef and dairy cattle ate, between them, some sixty-seven million tons of feed grains in 1974. Almost all of that grain could have been eaten by people.

A word about *feed* versus *food* grains. Feed grains—mainly corn and sorghum in this country—are grains which are grown to feed animals. Food grains are grown to feed people. Feed grains *can* be eaten by people, and in some countries *are* eaten by people, but they are not planted for that purpose. Almost all the corn you see growing in fields in the fall, for instance,

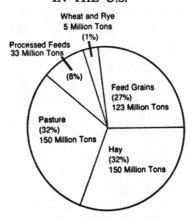

WHAT ANIMALS ACTUALLY EAT IN THE U.S.

Wheat and Rye
5 Million Tons
(1%)

Processed Feeds
33 Million Tons
(8%)

Feed Grains
(27%)
123 Million Tons

Pasture
(32%)
150 Million Tons

Hay
(32%)
150 Million Tons

Source: Anderson, 1975.

is what is popularly known as "field corn," varieties of which are eaten by animals, not people. Humans can eat field corn, particularly if it's picked young, but it is not as tasty as "sweet corn," varieties of which are planted to feed people. Sorghum, in this country, is used exclusively to feed animals with the exception of some sweet varieties that are turned into a pancake-type syrup. In some parts of South America and Africa, though, sorghum is the principal food grain. The chief food grains in our country are wheat and rye, almost all of which are eaten by people. Animals get some of them, though; according to Anderson's figures, livestock ate five million tons of wheat and rye in 1974. The United States feeds more grain to animals than any other country, although the practice is also followed in Western Europe, Russia, and Japan.

Beef cattle get most of the feed grain they consume in what is called the feedlot, a commercial enterprise that exists solely to bring beef cattle to marketable weight at the highest price possible. For every pound of beef a cow puts on in a feedlot, it consumes anywhere from seven to sixteen pounds of grain, depending on how much the animal weighed when it was purchased, how long it stays in the feedlot, and other factors. Frances Moore Lappé, author of *Diet for a Small Planet*, claims the figure hovers near the upper limit. The National Livestock and Meat Board, an industry association representing beef growers, claims it's closer to the lower limit. Whether it's seven pounds of grain to one pound of beef, or a higher figure, the fact remains that the conversion rate in the feedlot is inefficient, not by U.S. economic standards, but in terms of world food needs. And it is this inefficient conversion rate of grain to beef that excites many environmentalists, who claim we can no longer afford to raise grain-fed beef in a world where grain is in short supply.

Well, then, why not simply let cattle eat whatever they ate before the rise of the feedlot? That institution is, after all, a fairly recent innovation and a U.S. one to boot. In most other countries, including the big meat exporters Australia, New Zealand, and Argentina, cattle eat whatever is available on open rangeland; there are few, if any, feedlots.

The reason for the existence of the feedlot in the United States today is simple: Cattle fattened on grain before slaughter bring more money. Bart P. Cardon, head of Arizona Feeds, one of the nation's largest feedlot operations, told a group of agricultural scientists recently that grain-finished cattle receive higher grades under the U.S. Department of Agriculture's meat grading system. The higher the grade, the higher the price. The chief reason for the higher grades is that grain-fed beef has more fat *within* the flesh—the condition known as marbling—which makes it juicier and more tender. Grain-fed beef also shrinks less in processing because it has less moisture, making it worth more after butchering. A grain-fed animal weighing one thousand pounds would be worth 40.6 cents a pound on the hoof, Cardon noted. An animal of comparable weight fed on pasture—no grain—would be worth 17.5 cents per pound *less* than the grain-fed animal. It's possible, he conceded, to fatten a steer to a better grade on nothing but grass, but the feat is rare in his estimation. Not only that, but a grass-fed steer, even if it weighs the same as a grain-fed steer and has similar marbling, shrinks more when dressed because of moisture loss, thus giving the farmer and feedlot operator a smaller return.

"And *that's* why we use grain," Cardon summed it up.

Another reason for current grain-feeding practices in the United States is—or was—the low price of grain. The three decades after the end of World War II were the heady days of grain surpluses, when it seemed as if all the corn, wheat, and other food and feed grains produced by our efficient farmers would never be exhausted. Corn sold for $1.20 a bushel right up until 1971, a price which made it economical to feed corn to cattle, not only for a short stay in the feedlot, but for longer and longer periods. The day of huge grain surpluses ended suddenly in 1972, along with cheap grain. In 1977, as I write this, the price of corn is $1.89 a bushel. It has been much higher in recent periods (bumper crops lowered prices in 1977) and most experts think grains will never again be as cheap as they were in the postwar years. As a result, cows are eating less today, although the record grain crops of 1976 and 1977 upped grain consumption recently. In 1974–75, cattle ate some 53 million tons of grain; in 1976–77, the figure was expected to be 63 million tons. Another development that affects the amount of grain beef cattle are fed is the recent change in the U.S. Department of Agriculture's meat grading standards. The change upgrades beef with slightly less marbling, which means that slightly less grain will be needed to produce the desired marbling. All in all, the trend to grain is down. If you consider the present-day beef cow's diet over its entire lifetime, not just the period in the feedlot, the National Livestock and Meat Board claims it eats just three pounds of grain for each pound of meat it produces.

What do cows eat when they don't eat grain? Forage. Forage is an all-encompassing term that takes in native grasses that still grow wild on the western plains; cultivated grasses, such as Bermuda grass; cultivated forage legumes, including alfalfa; and the hay and silage made from these plants. Hay consists of dried grass or forage legumes; silage is a fermented product made by placing grasses or forage legumes within the airtight confines of a silo. Both hay and silage are fed to cattle during periods when green forage is not available. Cows can live perfectly well on an all-forage diet. In fact, that's just the way cattle did live until this century in the United States and still do live in most areas of the world. In the United States, forage-fed cattle are usually given an inexpensive synthetic form of nonprotein nitrogen such as urea, which enables them more easily to synthesize protein from the cellulose in forage. A bull referred to by the prosaic name of Number 1486 lived for twelve years on a protein-free diet that included urea at the U.S. Department of Agriculture's Beltsville, Maryland research farm. Although he grew a little more slowly, he eventually attained the same adult weight as protein-fed bulls. Number 1486 sired numerous calves, some of them with heifers fed a protein-free diet.

Cows flourish on a forage diet because of their remarkable digestive system. Animals can be divided into three main groups on the basis of their method of turning food into bodily nutrients. Man, pigs, chickens, dogs, cats, and a host of other creatures have a single stomach. Digestion is carried on principally in the small intestine, where various enzymes help break down foods into amino acids and other food components. Single-stomached beasts need foods with easily assimilated nutrients. Cows and other ruminant plant eaters, a group which includes such diverse animals as deer, buffalo, and sheep, have four stomachs. Digestion is carried on principally in the first stomach, the rumen, which has an array of bacteria that break down foods into their components. So efficient is this system that ruminants can extract nutrients from the fibrous material called cellulose that makes up the bulk of forage plants. Single-stomached creatures can't digest cellulose. An intermediate group of beasts composed of plant-eating animal families such as horses and rabbits possesses digestive system bacteria that break down foods to some extent, but they are unable to eat all of the forages that nourish the ruminants.

Forage plants are hardy plants, far hardier than such crops as grain and food legumes, and they can survive in areas where food crops do poorly. About 60 percent of the world's land area and about 63 percent of the United States is classified as grazing or range land. That amounts to about 1.2 billion acres in the United States alone. At present, only about 20 percent of our grazing land is being used to full capacity. For the most part, forage land cannot produce crops consistently without permanent damage to the soil. A large part of the world's land, in other words, is suitable *only* for plants that can be used to nourish ruminants, including the cow. Other foraging domesticated animals are sheep, goats, and water buffalo. Sheep and goats are the world's most efficient foragers, capable of living on marginal lands which could not support cattle. Some of these animals forage too well; numerous islands and some areas of the Middle East and Europe have been eaten almost bare by sheep and goats.

In most areas of the world, ruminants eat native plants on rangelands that are not planted. The homesteaders who settled the American West didn't *plant* forage; their cattle and sheep ate what was available on the vast western rangelands. In areas of the West, natural rangeland continues to support animals. Dr. Glenn W. Burton, a U.S. Department of Agriculture research geneticist who specialized in forages, grew up in western Nebraska, where he still owns a ranch. About half of the ranch has never been planted and his cattle eat native grasses. In the eastern United States, on the other hand, pasture for cattle and sheep is planted in a variety of forages, most of them imports. The most popular planted forage in the

U.S. is alfalfa, a legume that originated in the Middle East. Other popular forage crops are white and red clover, both legumes, and grasses such as orchard grass, bromegrass, and Bermuda grass.

Until recently, not much monetary support was available for forage research, but that situation is changing as a result of the world food shortage. Today a number of research programs are under way in this country that will, one hopes, improve forages to the point where cattle will be able to grow on them almost as fast as they do on grain. Some recent successes indicate that such a goal is possible. Glenn Burton's USDA project at Tifton, Georgia, the largest forage-breeding program in the country, has released a variety of coastal Bermuda grass that yields twice as much dry matter as common varieties of the popular forage. In trials in three southeastern states, it produced over one hundred pounds more weight gain per acre than common Bermuda grass. When the new variety and the common Bermuda grasses are well fertilized, the improved grass results in an even bigger weight gain. About ten million acres have been planted to coastal Bermuda grass in the southern United States and Burton estimates that the super forage contributes one billion pounds of beef per year more than would be produced with a similar acreage of common Bermuda grass.

Like a growing number of agronomists, Burton thinks U.S. beef cattle will soon be eating more forage. Someday, he predicts, they may reach market on nothing but forage. "We've got to recognize that the world is not going to let us feed grain to animals, even if we can afford it," says Burton. "I think it's a logical assumption that most U.S. cattle will be finished on forage someday."

Ruminants have still another dietary advantage over animals with simple stomachs: They can digest a number of cellulose waste products. Wastes are already being fed to cattle on U.S. farms. Corn silage, a popular cattle feed, is made by grinding up the whole corn plant—leaves, stems, cobs, and grain—and fermenting it in a silo. But there are numerous other wastes that the cow would regard with at least mild enthusiasm. Dr. Terry J. Klopfenstein of the University of Nebraska has compiled a list of suitable cellulose waste products that includes a variety of food processing wastes (bagasse, rice straw, corn gluten meal, grain wastes from the brewery and distillery industries), fruit and vegetable wastes (stems, leaves, and cobs), and paper industry by-products (waste paper, sawdust). Even straw, which consists of the inedible stalks of various grains after threshing, can be eaten by ruminant animals if subjected to fermentation. Chemical treatment aids the digestibility of other cellulose crops, Klopfenstein found. In one experiment he carried out, calves eating corncobs treated with sodium hydroxide gained three times as much weight as calves eating untreated cobs.

The cellulose waste with the biggest potential for use as cattle feed is livestock waste, which contains both organic and inorganic nutrients. An estimated 1.6 billion tons of animal waste is produced each year in the United States alone. Dr. W. B. Anthony and a group at Auburn University in Auburn, Alabama, have shown that mixing fresh cattle manure with grain and hay produces a feed that is almost as nutritious as a conventional mixture of grain and hay. When cattle are fed a mixture of hay, grain, and manure that has been converted to a kind of silage the Auburn group calls wastelage, the results are even better. Cattle in some wastelage tests actually gained more than their counterparts on a regular diet. But is manure-fed beef palatable? The Auburn researchers killed the young cattle used in one wastelage experiment and served steaks taken from the carcasses to some sixty guests at a dinner. Steaks taken from conventionally fed beef were also available. The diners couldn't tell the difference between manure-fed beef and grain-fed beef. An analysis of the two types of beef with an instrument called a tenderometer produced the same results.

Poultry litter, a protein-rich substance made up of a rather unlovely mixture of poultry excrement, uneaten feed, feathers, and a variety of cellulose-type beddings used in indoor poultry farms is another nutritious (about 20 percent protein) source of cattle feed. Dr. J. P. Fontenot of Virginia Polytechnic Institute and State University in Blacksburg has obtained good results by adding up to 27 percent poultry litter to corn and water and putting the mixture in silos to ferment. In one experiment, cattle liked some of his litter silage so much they ate it faster than they did a regulation corn-soybean silage. Utilization studies show cattle do just about as well on the poultry litter diet as they do on conventional diets. Sheep also fared well on the Fontenot diet. When the Virginia Polytechnic researcher upped the percentage of poultry litter to 50 percent, however, cattle ate it in lesser amounts.

The work of Fontenot, Anthony, and other researchers on animal wastes is so promising that cattle manure and poultry litter would probably be in the cattle diet today if it weren't for a major stumbling block. The U.S. Food and Drug Administration does not sanction the use of either cattle manure or poultry litter in feeds because of the possible presence of bacteria, metal residues, drugs, or pesticides in animal wastes. To find out if one harmful bacterium that is responsible for many cases of human food poisoning, Salmonella, could survive during the production of wastelage, Anthony inoculated some feed-manure mixtures with the organism and put them in a simulated silo. Under the usual temperature conditions that exist in silos, he discovered, Salmonella die. Fontenot got similar results with poultry litter. Further research on possible pathogens in animal wastes is needed but the work that has been done thus far is so encouraging that it looks as though manure-fed beef will be a reality in the future. If the prospect doesn't please you, remember that all that juicy beef finished at feedlots incorporates at least some manure. Cattle in the tight confines of a feedlot lie in their own wastes and lick themselves almost continually, transferring some manure into their bodies.

Manure-fed beef is still in the future, but some forage-fed beef, usually referred to as grass-fed beef, is already available. Actually, the term "grass-fed beef" is something of a misnomer. According to the American Meat Institute, *all* grass-fed beef in our supermarkets, with the exception of hamburger (some of which is made from grass-fed beef imported from abroad), comes from animals that have also eaten something other than grass, if only a product such as corn silage. The exact amount of grass-fed beef available in any one year is hard to estimate, because the percentage varies depending, among other things, on the size of the corn crop. What does grass-fed beef taste like? I've never knowingly eaten it, but meat experts say it tends to be less tender and juicy than grain-fed beef, mainly because of the lack of marbling. But according to "Ruminants as Food Producers," a booklet issued by the Council for Agricultural Science and Technology (CAST), beef with a low choice grade and "abundant marbling" fully equal to that of grain-fed beef has been produced experimentally in animals kept on an all-grass diet.

The experiment had a result not likely to please the cattle raiser, however: Meat from grass-fed animals weighed about 4.5 percent less than that from grain-fed animals of the same live weight, resulting in a thirty-dollar difference per animal in price. The difference in weight, explains CAST, is due to the greater size of the digestive tract in grass-fed beasts, the larger size being needed to handle the forage. Also, the grass-fed animals had to be fed about a month longer than the grain-fed specimens to reach the same weight. Nevertheless, CAST, which is made up of agronomists from all over the United States, hears the hoofbeats of forage-fed cattle on the trail. "Ruminants themselves can make the transition with ease," points out the CAST pamphlet. "The problems are with people—their nutrition, their preferences, their institutions and their economics."

Because of the cow's adaptability to a forage diet and wastes, agronomists and livestock experts agree that it is likely to remain around as long as man does. Consumption of beef, however, seems destined to decline, if only for economic reasons. Even beef cattle that live entirely on forage and wastes are expensive animals for reasons that lie in the nature of the beast itself. Cattle require lots of room (a minimum of five acres per animal for foraging), mature slowly, and bear only one calf every year. A cow is two years old before she bears her first calf. Almost any other farm animal can do better than the cow in growth and reproduction rates. Some research is underway to hurry up beef cattle's lethargic way of life—there are, for instance, attempts to induce cows to bear two calves instead of one—but it looks as though the cow will be an even more expensive animal to maintain in the future than it is today.

The dairy cow is in a stronger position than beef cattle because of its efficiency in producing milk. According to Wayne Anderson of *Feedstuffs*, the dairy cow produces 1.14 pounds of milk protein for every pound of grain and soy protein it eats, the best conversion ratio of any animal on the farm or ranch. Not only that, but the dairy cow usually produces a calf a year during its milk-producing life and, after it stops giving milk, can be turned into cheaper cuts of beef. Agricultural technologist Keith Barron of the Dow Chemical Company figures that the actual protein in low-fat milk (3.3 percent) costs about five dollars per pound, compared to eight dollars for boneless lean beef. Milk, of course, is also the starting point for cheese, another economical and nutritious form of animal protein. Like beef cattle, dairy cows can maintain themselves very well on an all-forage diet, but their milk production drops significantly. Nevertheless, the dairy cow, even more than beef cattle, seems assured of a place in our future.

What about the outlook for other popular U.S. meat animals? Wayne Anderson recently assessed the relative efficiency of all common U.S. meat animals with respect to their consumption of grain and soy protein versus their production of animal protein. After the dairy cow, he found the sheep the most efficient converter of protein (2.4 to 1), chiefly because most of their diet is forage. Then why is lamb so scarce and high-priced? The U.S. sheep industry, unfortunately, is beset with a number of problems that have nothing to do with the conversion efficiency of its main component, including vast land requirements, lack of labor (few young people want to be shepherds today), predator depredations, and an inexplicable but growing U.S. aversion to lamb. While beef consumption has been steadily rising in recent decades, lamb consumption has fallen off to where it is now only 2.0 pounds per capita each year, compared to 5.4 pounds per capita in 1920. Personally, I prefer a juicy lamb chop to a juicy steak any day, but the average U.S. consumer doesn't share my tastes. Sheep's minimal feed requirements, however, may possibly renew the popularity of these super foragers, which also supply us with valuable wool.

Other relatively efficient converters of feed protein to animal protein, according to Anderson, are the laying chicken and the broiler chicken, with the layer doing a little better than the broiler. Layers have a 3.0 to 1 conversion ratio, broilers a 3.4 to 1 ratio. If you consider reproductive capacity, the chicken also does very well, producing some twenty dozen offspring every year. But the chicken, like ourselves, is a single-stomached creature, requiring, under U.S. management techniques, a diet that consists almost exclusively of corn and soybeans. Research is underway into diets that include a higher percentage of waste products, but the chicken will probably always require a large proportion of protein that could be used by humans.

The turkey lags behind the chicken in efficiency, coming in with a 5 to 1 conversion ratio of feed protein to meat protein. It, too, eats mostly corn and soybeans.

The least efficient animal? The hog. Anderson terms the hog "the biggest competitor with man in terms of their use of grain and oilseeds." The hog's digestive system, as well as its other body systems, is so similar to man's that the animal is becoming more and more popular as a medical research subject. What's good for medical research is bad for economics, though, because the hog needs a diet much like our own. On modern U.S. hog farms, the hog's food consists largely of corn and soybeans, which are transformed from vegetable protein into pork protein at a conversion ratio of about 8 to 1. Per capita pork consumption has declined in the United States and many experts besides Anderson consider the hog the animal most "vulnerable" to the rising pressure to divert food grains to people. Hogs are efficient in one respect, however: reproduction. A good sow can produce litters of a dozen piglets twice a year.

Of course, we don't have to depend completely on traditional animals such as pigs, chickens, sheep, and cows to supply meat. Why not set our sights a little wider and consider other beasts that can be raised in large numbers—the buffalo, for instance? For,the first hundred years or so of our nation's history, the buffalo supplied a large part of the meat in the American diet, particularly in the West. The big, shaggy cattle were almost exterminated in the late nineteenth century but today they're making a comeback. There are at least thirty thousand buffalo in the United States and Canada, many of them in some three hundred commercial herds that supply buffalo meat to specialty stores. Buffalo spend most of their time foraging on open range, but those designated for the table are fed grain and hay for a brief period before slaughter. I've never eaten buffalo meat but fans claim it's richer and better-tasting than beef, yet just as tender. It seems unlikely that the buffalo could play a significant role in the U.S. diet again, however. Today buffalo need tremendous amounts of valuable rangeland and special fencing ("Buffalo are dangerous wild animals," warns a U.S. Department of Interior pamphlet), factors which make them more of a hobby than a business. Buffalo meat costs the consumer even more than beef. Those cheap buffalo of the nineteenth century were free-roaming creatures that cost nothing to maintain.

But large-scale management isn't the only answer to putting animal flesh on the table. An increasing number of U.S. families are taking a leaf out of their forebears' book and producing at least some of their own meat, milk, and dairy products. A good source of information on small- and medium-size livestock is the magazine *Countryside* (see Suggested Reading), which is directed to what it calls the "homesteader." By its definition, a homestead produces enough food for a family's sustenance but not enough for commercial sale. The editors say you can homestead on as little as an acre of land. According to *Countryside* and other knowledgeable sources, one of the easiest animals to raise for the family without much land is the rabbit. Domesticated breeds need little room, can subsist on grass, vegetable scraps, and roots (although many live largely on cheap rabbit pellets made out of alfalfa), breed prolifically, and reach eating size at an early age. A good breeder can produce four litters of six to eight young during the year. The young are at their best for eating just eight weeks after birth, although older rabbits can be used as roasters and stewers. Rabbit meat is delicious. I still remember the fried rabbit I ate as a child when my father returned from hunting trips. Why don't I eat rabbit today? My husband kept rabbits as pets when *he* was a child and the thought of eating a "bunny" makes him feel sick, a feeling shared by many other Americans.

If you can't face a meat dinner made from a pet, you might want to consider the opossum, this country's only marsupial. Although highly regarded for its flavorful, reddish-colored flesh in the South, it's not nearly so cuddly as the rabbit. The Possum Growers and Breeders Association (the correct name of the animal is opossum but almost everyone in the U.S. calls them possums), an organization with one hundred active members that has its headquarters in Clanton, Alabama, is trying to establish the wild opossum as a farm-bred animal via selective breeding. According to the president, Frank Clark, a captive-bred opossum weighs about thirty-five pounds compared to about five to twenty pounds for its wild brethren, and tastes "like roast pork." In the wild, opossums eat both vegetation and meat, but Clanton's captive animals receive a diet of high-protein concentrates like the kind fed to pigs and chickens. Over the phone, Clark was a little vague as to conversion ratios, but a few days later a package from Clanton arrived with a license plate reading "Eat More Possum," a certificate stating that I was a lifetime member of the Possum

Growers and Breeders Association, and a cookbook entitled *How to Cook a Good Possum.*

The cookbook has a number of interesting but cryptic recipes. One begins offhandedly, "Take one fat possum, cook until tender in a large pot . . ." So far I haven't been able to try out any of the recipes because the only way you can obtain a farm-bred opossum today is to buy one live from a member of the Possum Growers and Breeders Association.

For the present, at least, a much stronger case can be made for the home-raised dairy goat as a source of protein. Archaeological excavations show that goats and sheep were the first livestock domesticated by man, and these ancient beasts retain their popularity today in the Middle East and a number of other countries. And for very good reasons. Goats, in particular, are the world's best foragers, capable of subsisting on practically anything, although they perform better on grasses, alfalfa, and roots, supplemented with high-protein pellets for milking goats. They need only about an acre of land per animal. Their products? Principally milk and cheese, although some people do eat goat meat, which is known as chevon. A goat gives up to four quarts of milk a day for ten months a year for up to ten years. Some people swear by goat's milk. It tastes very much like cow's milk, but its slightly different properties make it more acceptable to some of the people who are allergic to cow's milk. In recipes, it is used just like cow's milk. Goat's milk makes good cheese, as the traditional Greek goat's milk cheese, *feta*, attests. Goat's milk has one drawback, however; it is higher in cholesterol than cow's milk, a factor of particular importance in adult diets.

Goats have never been terribly beloved in this beef-rearing country but their virtues are at last becoming recognized, particularly in agricultural areas with restricted land. When I visited several New Jersey agricultural fairs last year, the most popular animal being shown by far was the goat. The American Dairy Goat Association of Spindale, North Carolina, says goat registrations in this country rose from 8,144 in 1971 to 32,459 in 1976 (human membership in the association quadrupled in the same period). Does a goat belong in your backyard? Maybe. But remember that goats, like all animals (even dogs), offer problems. A fairly substantial shelter is mandatory, a milk goat must be milked daily, and the goat's appetite for just about any kind of food will lead it into your garden or that of your neighbor if you don't restrain it with a high fence or tether. Still, the advantages of these frisky beasts are such that I'm thinking of putting a goat on our two-acre plot—if I can just think of what to do with four quarts of milk a day.

11

FROM GARBAGE PAIL
TO DINNER TABLE

Little Miss Muffet
Sat on a tuffet
Eating of curds and whey.

—MOTHER GOOSE RHYME

THE TEST KITCHEN AT THE U.S. DEPARTMENT OF AGRICUL-
ture's Eastern Regional Research Laboratory in a suburb of
Philadelphia is so clean it looks as though no food had ever
been cooked there (a false impression, as I learned). A long,
narrow room lined with gleaming white appliances, it has a
big food preparation table, also immaculate, in the center.
On the day I visited the kitchen, there was a big bowl of
macaroni salad on the table, along with some small dishes.
Food technologist Jan Clark, who wore a white coat as pristine
as the room, spooned some of the salad into three dishes, one
for me, one for Virginia Holsinger, a chemist who had accompanied me to the test kitchen, and one for Robert L.
Miller, a program assistant who was conducting me on a tour
of the center. It was getting close to lunchtime and I finished
my salad rapidly. Clark gave me another bowl, which I also
finished.

"I like it," I said. "It tastes just like macaroni."

"To me, standard macaroni is more rubbery," said Clark

judiciously. "This is a little gritty." A taste panel, she added, had pronounced the same verdict, although all the panel members liked the macaroni.

What's so different about this particular macaroni? Apart from the possible grittiness, which I didn't notice, it's much more nutritious—almost as nutritious as milk casein, the much-used reference protein, which, theoretically at least, supplies enough essential amino acids to meet all our needs. Standard macaroni, which is made from durum wheat, has about 13 percent protein and a PER of only 0.8. The ERRL macaroni has 20 percent protein and a PER which ranges from 2.37 to 2.77, averaging 2.41. The source of the increased protein is whey, a by-product of cheese making that was once dumped into the nearest river or fed to animals. In fact, a lot of whey still is fed to animals or dumped into streams, even though it's capable of turning pasta into a food almost as nutritious as milk.

In its original state, explained Holsinger, whey isn't suitable for incorporation into macaroni or, for that matter, into any other processed food. Whey is the watery—93.5 percent liquid—substance left over after milk is coagulated in the cheese making process. The solid portion is the curds; the watery part is the whey. The curds are pressed into blocks or wheels to make cheese, squeezing out more whey. In the days of Little Miss Muffet, people ate both curds and whey, but whole liquid whey doesn't find much of a use today except as an animal feed on farms where cheese is produced. Almost all whey used today is either dried or condensed. To make a whey product that could be mixed into pasta, engineers at the ERRL heated cottage cheese whey, a procedure that makes the proteins coagulate, or clump together. When the cooled protein slurry was put through a centrifuge, a machine that removes moisture by rapid rotation, a solid cake with from 85 to 95 percent protein emerged. After drying, it can be mixed with durum wheat in a conventional "pasta press," a decided advantage if the macaroni is to be produced commercially.

One added point: Those taste panel members not only liked the whey-enriched pasta, they liked it better than protein-enriched pasta now on the market. These products depend on substances such as wheat germ and yeast for their high protein content.

The reason why the ERRL, one of four regional centers operated by the Agricultural Research Service, an arm of the U.S. Department of Agriculture, is giving all this attention to a whey product that can be put into pasta can be traced directly to the Water Quality Act of 1965. That law mandated a cleanup of the nation's waterways and included, among other provisions, a statute requiring the various states to set water quality standards. The standards limit the amount of BOD, or Biochemical Oxygen Demand, a waste can put on the water into which it's discharged. Whey and other food processing wastes exert a high BOD because of their nature as organic compounds. Small amounts of organics discharged into an unpolluted stream are readily oxidized by bacteria using oxygen already in the water, but large amounts cause trouble. They stimulate the bacteria to use oxygen faster than it can be replaced, leading to the smelly, opaque conditions characteristic of oxygen-depleted water. Fish and other normal forms of aquatic life can't exist in such waters.

Some large food processing companies treat their organic wastes, but small ones often can't afford the expensive treatment equipment. Instead, they pay a local sewage plant to do the job. Unfortunately, most community sewage plants also lack the equipment to handle large amounts of organics.

A solution to the problem that looks increasingly attractive is retrieving nutritious components from suitable food waste, an approach that not only gets rid of some of the waste but supplies a salable product. Not all food wastes, of course, are sources of protein, but a surprising number are. Some of the food by-products being suggested as possible protein sources are whey, the grain "mash" left over after distillery and brewery operations, and a wide range of animal by-products, including hair, blood, and even feathers. All these wastes include ample amounts of protein. One or two are suitable only for animals but research already indicates that most of them can serve as alternate protein sources for humans as well. Some of these by-products already form part of the diet of people in some areas of the world. Blood sausage, for instance, a traditional favorite in Eastern Europe, can be purchased in the U.S. in some specialty pork stores.

Of all these wastes, whey looks like the best candidate for protein retrieval in the United States. Not only does it have some 13 percent protein, 10 percent of which is the essential amino acid lysine (both figures are on a dry-weight basis), but an enormous quantity of whey is being produced today and there will be even more of it in the future. According to Virginia Holsinger, the United States produced some thirty billion pounds of whey in 1976. Consumption of dairy products in general is declining but consumption of cheese is rising, a trend dairy experts predict will continue for the foreseeable future. For each pound of cheese we produce, nine pounds of whey are left over. Some of it is already going into our diet. Large dairy firms, as well as a new industry, the whey processing industry, now process and sell about 55 percent of the nation's whey. It is marketed in a condensed form, as a dried whole whey powder, and as a variety of high-protein, desalted whey derivatives. A little more than half of it goes to animals. People get the rest, most of it in the form of dried products.

You never eat whey? Check your food cupboard and refrigerator. Unless you're a rather unusual consumer, you'll find at least one or two processed foods listing whey as an ingredient. In my own seriously depleted kitchen (I hadn't been to the market that week), I found whey listed on a package of hot dog buns and a package of dog-training treats, which neatly divides our whey consumption fifty-fifty between animals and people. As far as food products go, the major users of whey in the United States are the bakery and dairy industries. Cookies are particularly likely to contain whey. In these foods, whey is used almost exclusively for its functional properties, not for nutrition. Whey supplies a number of desirable functional qualities, including flavor intensification, color enhancement, thickening, adhesion, emulsification, and whipping capacity. As a whipping agent, whey exceeds soy, making the dairy waste a better egg replacer. In most foods, dried whey powder is a replacement for nonfat dry milk powder, which costs about five times as much.

The American consumer obviously eats a fair amount of dried whey, even if he or she doesn't know it, but more whey is consumed in Europe, largely in liquid form. Whey food is an old story there. The most famous consumer is Little Miss Muffet of Mother Goose fame, who, as you'll remember, was eating curds and whey when frightened by a spider. I'm indebted to Harold W. Rossmore of Wayne State University for the information that Miss Muffet may have been a real person, the daughter of a sixteenth-century figure, Sir Thomas Muffet. An expert on spiders and nutrition (people were more versatile in those days), Sir Thomas advocated the use of fresh cheese which, in those days, might well translate into curds and whey, the freshest form of cheese possible. He had one daughter. Whether Miss Muffet was real or not, though, she certainly wasn't alone in her liking for whey. In a paper he delivered at a whey conference, H. C. Trelogan of the U.S. Department of Agriculture's Statistical Reporting Service noted that two famous Englishmen, Samuel Pepys and Sir Walter Scott, drank whey. ". . . And thence to the Whayhouse and drank a great deal of Whay," runs an entry in Pepys' diary dated June 10, 1663. Scott drank whey when visiting a relative, not, as he reported in a letter, because of its nutritive value but because it was brought to his bedside every morning by a pretty dairy maid. During this period, whey was easily obtainable in England even in cities, both in whey houses or from farmer-vendors who sold whey in the street.

Although many whey drinkers seem to have enjoyed the beverage, whey was drunk more for its reputed therapeutic benefit than for its taste. Holsinger found that the earliest mention of whey was in the fifth century B.C., when Hippocrates, the Greek physician, prescribed it for an assortment of human ills. Physicians continued to prescribe whey through the centuries. "There are few fluids more salutary, and better adapted to prevent and cure the diseases of the human body than whey," wrote Dr. Frederick Hoffman, physician to the King of Prussia, in 1761. In the mid-nineteenth century, there were over four hundred whey houses in Western Europe which offered something called a whey cure. During a cure, the ingestion of up to 1,500 grams (52.2 ounces) of whey per day was prescribed for ailments ranging from arthritis to liver complaints. At Ischl, a famous Austrian spa, the whey cure took the form of a whey bath, which was claimed not only to calm the nerves but to soften the skin. It was particularly popular with women. As late as the 1940's, some European spas were still offering a whey cure.

Whey cures are a thing of the past but whey is still being imbibed in large quantities in Europe, at least partly for its therapeutic benefits. The most popular whey beverage in Western Europe is Rivella, according to Holsinger. Introduced in Switzerland in 1952, it is a clear, carbonated drink that contains an infusion of "alpine herbs." Other whey-based drinks currently being sold in Europe include whey champagne and kwas, both fermented, nonalcoholic Polish products, and Bodrost, a Russian fermented drink that tastes something like beer but has less than the usual amount of alcohol. A number of whey beerlike beverages with a normal alcohol content were concocted and drunk in Germany during World War II, when supplies of standard beer ingredients were short. Whey proved to be a good beer substitute because it contains a material similar to beer wort, the sugar solution obtained from malt that is fermented to make beer. After the war, however, whey beer was quickly replaced by traditional beer.

From the standpoint of protein, there's just one drawback to these new whey beverages: They contain little protein, having undergone a process called deproteinization in which most of their protein is removed. In a typical whey soft drink sold in Poland, deproteinization removes about 63 percent of

NUTRIENTS IN DRIED WHEY AND
NONFAT DRIED MILK

NUTRIENT	APPROXIMATE CONTENT	
	WHEY PERCENT	NONFAT DRY MILK PERCENT
Protein	13.0	36.0
Fat	1.0	1.0
Ash	8.0	8.0
Water	5.0	4.0
Lactose	73.0	51.0

Source: D. A. Vaughan, 1970, 1977.

the protein, and the addition of fruit juice removes still more. The final product has only about three to four grams (.14 ounce) of protein per liter (about a quart). Deproteinization is used primarily to maximize keeping quality. Whey drinks may not be particularly nutritious, but the protein obtainable from deproteinization is usable for whey protein concentrates, giving whey beverage processing a nutritional aspect. And deproteinization *does* convert waste to salable products. Thus far, no whey-based drink has caught on in the United States, although Rivella has been test marketed here. Virginia Holsinger thinks the reason why whey drinks are so popular in Europe is that soft drinks similar to ours are not available in many countries. In the home of Coca-Cola, whey-based drinks have a hard time capturing a market.

But there must be something we can do with the 45 percent of our whey that is literally going down the drain. Ever since the Water Quality Act was passed in 1965, a number of private and public organizations, including the USDA's Eastern Regional Research Laboratory, have been looking for new uses for whey, particularly nutritional uses. On the face of it, converting whey into a nutritional additive seems like a simple proposition. Why not just drop some dried whey in low-protein foods and watch their PER rise? But putting significant amounts of whole whey in food creates problems. Although dried whey is used primarily as a substitute for nonfat dry milk in processed foods, there are important differences in the two products. Dried whey has less protein than nonfat dry milk (13 percent compared to 36 percent) and smaller percentages of the essential amino acids. On the other hand, whey has more lactose, or milk sugar, and more potassium and sodium salts than milk. Nonfat dry milk has about 52 percent lactose, whey about 74 percent lactose. The main component of whey is lactose; there is almost six times as much lactose in whey as there is protein.

Large segments of the populations of Asia and Africa cannot digest lactose because they lack an enzyme called lactase that breaks down lactose into its component sugars. When lactose-intolerant people ingest more than a small amount of lactose, they develop diarrhea, cramps, flatulence, and other digestive complaints. The same problem arises with milk and milk products but since whey contains more lactose than milk, smaller amounts of whey will produce the symptoms. From 7 to 15 percent of the population of the United States is estimated to be lactose intolerant, mostly blacks, American Indians, and Orientals. Oddly enough, the condition does not usually develop until a child reaches school age. Some animals, including rats and chickens, are also lactose-intolerant. Because rats can't digest whole whey, no PER exists for it, only for low-lactose whey derivatives.

Older cats often develop lactose intolerance, so watch your aging tabby or tom's reactions carefully if you feed it milk. Diarrhea after drinking milk is usually a sign of lactose intolerance.

One way to get around the lactose problem and improve the quality of whey at the same time is to extract the proteins and leave the sugars and salts behind, much the way soybean processors do in making soy concentrates and isolates. A number of new processes exist to do exactly that. Essentially, they either separate the high-protein fraction of whey from the sugar and salt fractions, or concentrate the protein by removing the sugars and salts. In both cases, the end product is spray-dried to a powder that not only has little lactose and salt but a protein content of up to 95 percent. The ERRL has developed a form of high-protein whey concentrate that has a PER of 3.1 or 3.2, which exceeds that of any product except the hen's egg. It isn't on the market, but a number of whey concentrates developed by private firms are being sold to commercial users. All are designed for functional use but some of them will undoubtedly be used to augment the nutritive value of foods in the future. In one study conducted at the ERRL, researchers found that some forms of whey concentrate do not depress volume in bread, even when used at up to 6 percent of bread content.

Another way to deal with lactose is to take it out of milk before it is made into cheese. An ERRL staff member, Eugene E. Guy, has devised a method to remove lactose from milk by adding the lactase enzyme to milk and heating it, then keeping the beverage hot for a few hours. The process breaks down lactose into the sugars glucose and galactose in the same way the enzyme would if it were naturally present in the digestive tract. The milk products that emerge from the treatment, including whey, are much sweeter than usual because glucose and galactose taste sweeter than lactose, but the effect isn't necessarily a drawback. In some products, sweetness is a plus. Guy found, for instance, that when lactase-treated whey is used in ice cream, it reduces the amount of sugar needed.

But even if the lactose is removed from whey, a problem still exists. What do you do with all that lactose? Some is used today, chiefly by the baby food industry, but it's a mere drop in the whey bucket.

No one really has a solution to the excess lactose problem at this point but some intriguing new uses for whole whey and whey protein have been developed, some of them by the ERRL. The Philadelphia research center is particularly interested in what the dairy industry calls acid whey, a by-product of making cottage cheese and cream cheese. About one-fourth of the whey produced in the United States is acid whey. All the rest is sweet whey, a by-product of making all other cheeses. The big difference between the two by-products is that acid whey contains more lactic acid, a substance which gives acid whey a sharp, acid taste and makes it difficult to dry. Until very recently, almost all acid whey was simply thrown out. In the last few years, the ERRL has not only figured out a way to dry whole acid whey, but it has found some products to put it and its derivatives in. One of them is the macaroni I enjoyed in the test kitchen. It incorporates the 3.1 to 3.2 PER protein concentrate mentioned earlier, which is extracted from cottage cheese whey. The creators of the concentrate, engineers Howard I. Sinnamon, Edwin F. Schoppet, and Curtis Panzer, believe it will be usable in a wide range of pasta products, as well as in some breakfast foods and snacks.

Another vast potential market for acid whey targeted by the ERRL is the soft drink industry. If ever a product needed fortification, it's this one; soft drinks, which have a per capita use of about one can or bottle per day in the United States, have *no* protein—just water, sugar, flavoring, and, if they're carbonated, carbon dioxide. In an experiment Virginia Holsinger directed, various flavors of carbonated and noncarbonated soft drinks were fortified with a form of acid whey concentrate. When the drinks had 1 percent protein, even experienced judges couldn't tell the difference between some fortified beverages and the nonfortified ones. A raspberry noncarbonated drink including whey got even higher marks than the raspberry control drink. The cost of putting 1 percent whey protein in a soft drink would be on the order of about three-fourths of a cent for an eight ounce bottle. Holsinger tried making beverages with 3 percent protein, which would bring the soft drinks in the range of milk, but ran into flavor problems. Someday, however, she predicts, fortified soft drinks as nutritious as milk may be feasible. Meanwhile, as she puts it, "Any protein at all included in a soft drink represents a nutritional improvement."

The ERRL has also come up with several new uses for whole sweet whey. One is a snack spread with a mild cheesy taste and a butterlike consistency made from cream and whole sweet whey. It has 5.6 percent protein. The spread is designed primarily for the U.S. market but another new sweet whey product is aimed solely at children in developing countries. Eight million pounds of a whey-soy beverage are now being consumed in five countries under the Food for Peace program operated by the U.S. Agency for International Development (AID). Until a few years ago, nonfat dry milk was the beverage distributed overseas by AID, but in 1972 supplies became too scarce and expensive for continued use in the program. To find a substitute, AID turned to the U.S. Department of Agriculture. The ERRL came up with powdered whey-soy milk, a variation of a mixture that had been under development at the research center. The new beverage has a number of special attractions, including a low salt content. As it is constituted now, whey-soy milk consists of 42 percent dry whole whey and 37 percent soy flour, plus corn syrup solids and soybean oil. The powder itself has a protein content of 21.2 percent; an eight ounce glass has about 7 grams (.25 ounce) of protein. The PER is 2.1, which makes it almost equal to milk casein at 2.5.

Children in developing countries don't have too many choices of beverage, but taste tests carried out in six countries before the product was distributed showed the whey-soy milk was well liked almost everywhere. Lactose intolerance? No symptoms of it were reported. The last finding isn't too surprising, since pre-school age children seldom show signs of lactose intolerance. If problems do develop, though, the ERRL has a lactase-treated whey-soy formula available. In a taste test, it got even higher scores than the standard drink.

They may love whey-soy drink overseas, but it's a different story here according to Holsinger, who finds the same objections to the soy flavor that soy researchers at the USDA's Northern Regional Research Laboratory in Peoria, Illinois, find (see Chapter 2). "Our taste panels just don't like the soy flavor," she said. Actually, whey-soy drink isn't really that bad. Holsinger gave me a tiny paper cup of the beverage, which looks almost as white and opaque as milk. I drank it, thought it tasted rather like Loma Linda's Soyagen, which I like, and drank another cup. Because of the general aversion to the soy flavor, the best market for the drink here may be the health food store, where a PER of 2.1 may arouse interest. Thus far, though, the drink is not on the market anywhere in this country.

The ERRL isn't the only research organization that has found new uses for whey. You could easily get through the day on some of the whey products that have been concocted in various laboratories. For a light breakfast, you might want to try O-way, a 1 percent protein liquid meal developed at Michigan State University that includes four parts whey to one part orange juice. Another breakfast choice is a mixture of whey with grape juice and passion fruit juice dreamed up at the University of Arizona. A taste panel gave it a very favorable rating. At lunchtime, a soup made from tomato juice and whey looks like a good choice, especially with some crackers spread with a mixture composed of heat-coagulated whey protein and chives. The heat-coagulated whey substance, which was formulated by Pavel Jelen of the University of Alberta in Canada, can be used as a meat extender in your dinner sausage, too. A taste panel liked Jelen's whey-extended sausages considerably better than ordinary sausage. A baked potato with imitation sour cream made from whey should go well with the sausage. Before you eat your entree, though, you should try a very special whey cocktail—an 11 percent whey sherry invented by Father E. R. Engel of Palmer, Alaska. With, of course, whey-enriched rolls. Dessert might be a new whey fruit sherbet made with acid whey. And if you have a real sweet tooth, a bar of 5 percent protein whey marzipan made with a high-protein concentrate might be worked in during the day.

Only one of these products, fruit sherbet, is presently available to consumers, although the sour cream was on the market at one time. Why isn't nutritious whey used in more foods in larger amounts? One reason, argue many whey experts, is whey's *name*. Because it was used exclusively as an animal feed in this country until the last few decades, people associate whey with animal feeds, not human foods. D. R. Braatz of the Consolidated Badger Company in Shawano, Wisconsin, a whey processing firm, drew a parallel between whey and "skim milk," now known as nonfat dry milk, at a whey conference recently. It seems hard to believe now but, as Braatz pointed out, skim milk—sorry, nonfat dry milk—was once fed almost exclusively to animals. In 1944, the name was changed to nonfat dry milk and ten years later human usage of the product had increased from 500 million pounds to 1.3 billion pounds. Ten years after that, human usage was up to 2.1 billion pounds. Nonfat dry milk is still going strong. Could the same thing happen to whey? Whey processors like to think so.

Under any name, though, whey is too good to feed primarily to animals and definitely too good to become a pollutant of our waterways. We'll be eating and drinking more whey in some form.

The same is undoubtedly true of some other food wastes. The biggest group of these products is made up of by-products of the meat and poultry industry. Many of these by-products already find a market. Besides the familiar cuts of meat we see in the supermarket or butcher shop, the meat industry also provides the raw material for a number of edible by-products, as well as for many pharmaceuticals and inedible by-products such as leather. Many of the edible by-products end up in animal feeds. Some blood is used to feed animals, as are some feathers—yes, feathers—and some of the packinghouse wastes known collectively as offal. Much of our animal waste, however, is not used for anything and is simply discarded, creating a pollution problem in waterways. With more stringent water pollution laws now in effect, the meat and poultry industries, like the dairy industry, are looking for additional ways to use edible meat by-products.

One promising new market being investigated by scientists is foods. A surprisingly large number of these wastes, it turns out, are good not only for animals but for people.

Take, for example, the bits of meat and fat that cling to the bones after butchers slice off chops, steaks, and other cuts. Dr. R. A. Field, a professor of food science at the University of Wyoming in Laramie, estimates that some 2,097,757 metric tons of this meat are wasted every year as a human food source, although most of it is perfectly edible. Machinery introduced within recent years makes it possible to debone mechanically both meat and poultry. The protein content of various mechanically deboned cuts of red meat has been assessed in research studies at 9.0 to 17.57 percent and PER's run as high as 2.85. The protein content of mechanically deboned poultry parts ranges from 9 to 14 percent. Some mechanically deboned poultry is already being used in processed foods and pet foods, but no mechanically deboned red meat is permitted under FDA regulations. New rules have been proposed by the U.S. Department of Agriculture, however, and mechanically deboned meat may soon be allowed in food products ranging from sausage to spaghetti sauce, meaning that we'll be eating more of the meat that now goes to waste.

Another meat waste product that is slated to play a bigger role in our diet before long is offal. Offal consists of internal organs such as the spleen, stomach, kidney, trachea, bladder, and brain, as well as body parts such as pig snouts and beef udders. Some of these items are already in processed foods. The first time I read "pork snouts" on a package of lunch meat, it was something of a shock, but if you're a meat eater today, you have to take the snouts with the chops. Actually, snouts are not too hard to take, since they are unrecognizable in processed foods. The various organs and parts being suggested for wider human utilization will probably be similarly disguised. In a research project directed by Floyd C. Olson of Oscar Mayer and Company, and Ezra Levin of the VioBin Corporation, numerous organs and parts of both cattle and hogs were converted into a high-protein concentrate. One 1,271-pound steer yielded 28 pounds of concentrate. Suitable for augmenting the protein content of a wide variety of foods, the concentrate had from 78 to 90 percent protein and a PER that hovered around the 1.70 to 1.90 level, depending on the organs involved.

The organs with the best amino acid profiles? Hog kidney and brains, and beef esophagus, spleen, and intestines.

If offal makes you feel a little queasy, blood will probably have an even worse effect, but blood is a highly nutritious animal waste. Dr. F. W. Knapp of the University of Florida's Institute of Food and Agricultural Sciences estimates that our livestock industry produces up to two million pounds of blood each year. A single animal contributes about three and one-half gallons. Some blood is dried for use in animal feeds, but the major portion is simply disposed of through municipal or private sewage systems—at considerable expense. Whole bovine blood contains about 17 percent protein distributed in two fractions, a liquid fraction known as plasma and a solid one composed of the red and white blood cells. High-protein concentrates with from 70 to 90 percent protein have been made from both fractions. Work at the ERRL indicates that both whole blood and the concentrates are low in the essential amino acid isoleucine but plasma contains considerably more than either whole blood or the solids fraction. PER's as high as 2.83 have been reported for plasma concentrates, which makes them one of the better protein supplements. The PER of whole blood is only 1.79 but when it was combined with wheat in an experiment at the ERRL, the blood-wheat mixture yielded a PER of 2.1, which is higher than either wheat or blood separately. The finding indicates whole blood may make a good supplement for the low-protein grains.

Of course, there's another problem with blood aside from its low level of isoleucine: the color. Blood is, well, red. Awfully red. And while blood red is a desirable color in meat, it is a distinct drawback in bakery goods and many other products suitable for supplementation. The color, which is contained in the heme pigment, isn't as big a problem as it may seem, however. Heme is much reduced in the concentrates, which have a light tan color, and it can be removed completely by a fairly simple procedure. The resulting product is a white, odorless powder. Dr. Knapp and a group at the University of Florida formulated an imitation cheese out of a white blood concentrate, a tan concentrate, and whey. It looked and tasted much like cheddar cheese and had 18.9 percent protein (real cheddar has about 23 percent protein). Collecting blood offers more problems than decoloring it at this point. Only beef blood is considered edible in this country because of the way in which it is collected and even it must be drawn and stored under certain conditions to make it suitable for human use.

A number of other animal waste products also yield considerable amounts of protein, although none of them is as balanced as the protein from blood, organs, and mechanically deboned meat. Collagen is one such protein source. The major protein of skin, bone, and connective tissue, collagen is found

DIETARY PROTEIN VALUE OF COLLAGEN	
ESSENTIAL AMINO ACIDS	AMINO ACID LEVELS (%)
Tryptophan	0.0
Phenylalanine	2.4
Lysine	4.0
Threonine	2.3
Methionine	1.0
Leucine	3.7
Isoleucine	1.9
Valine	2.5
Cystine	0.0

Source: R. A. Whitmore, H. W. Jones, W. Windus, and J. Naghski, 1969.

in great abundance in beef and pork hides. Some collagen is used for products such as gelatin, sausage casings, and the new liquid protein diets, the major component of which is collagen, but much of our collagen goes to waste. Collagen is a prime example of an unbalanced source of protein. It has *no* tryptophan and *no* cystine and very little methionine. In fact, it's low in most of the essential amino acids. If you feed a rat nothing but collagen as its source of protein, it will soon die, making the PER of collagen zero. Why bother with a protein with a zero PER? Collagen is a very digestible protein food with texturizing and binding properties that would make it a desirable ingredient in foods such as meat extenders and imitation meat. But take a word of warning from the rats, which have the same type of digestive system as ours, and don't make the popular liquid diets your only form of protein. In spite of the fact that they are sold without a prescription, they should be used only under a doctor's care.

Animal hair is another unbalanced protein that may nevertheless yield valuable components for processed foods. At the ERRL, investigators have isolated a 90 to 92 percent white odorless powder from the hair protein, keratin, obtained from tannery wastes. In spite of its high protein content, the substance contains only small amounts of all eight essential amino acids. Rats that ate it along with soy protein and a lysine supplement gained less weight than they did on soy alone, giving the soy-hair-lysine mixture a low PER of 1.69. Digestibility was low, too. Work at the ERRL is continuing to determine whether further treatment could make hair more digestible, but meanwhile, J. W. Cooper, one of the members of the team that worked on the hair protein powder, has come up with a treatment that makes hair waste suitable as an animal feed, if not a human food. It involves adding activated sludge, a substance used in treating sewage, to liquid hair waste and letting the bacteria in the sludge acclimate themselves. Before long, the bacteria eat the hair, multiply, and produce a pale brown product with over 30 percent protein and a considerably better amino acid profile than the original sludge. At the same time, the tannery waste is clarified to the point where it can be introduced into streams without raising the Biochemical Oxygen Demand (BOD). Rats thrive on the protein-rich substance produced by the bacteria and Cooper thinks chickens will, too.

One form of keratin is already being eaten by chickens and it may soon be in our diet, too. A ground, heat-treated meal made from poultry feathers has been a component of poultry diets for a decade. The meal's unbalanced amino acid profile—it's low in methionine, histidine, and lysine—limits its use and it accounts for only a fraction of the 550 million pounds of feathers (an average chicken yields two to three ounces) produced every year in the continental United States. What happens to the rest of those feathers? They go into sewers or incinerators. Scientists have been trying to develop other food uses for feathers and in 1973, an Atlanta scientist, Dr. William D. Goodwin, came up with a patented process by which a white, feather-protein isolate can be produced. It has 96 percent protein and is highly digestible, even though it's low in the same three essential amino acids as feather meal. Further work with the process by Dr. John Paul Cherry, then with the University of Georgia but now a member of the USDA's Southern Regional Research Laboratory in New Orleans (see Chapter 7), indicates that isolates with various characteristics may be able to play a role in human foods.

Cherry put the isolate in sugar cookies at up to 15 percent of the contents, and submitted them to a test panel, which gave them high marks. "We had no complaints and nice remarks," he said.

Animal wastes are the major source of waste protein in the United States, but not the only one. Probably the most successful industrial effort to transform a waste into a nutritional product is the conversion of the grain by-products of the distilling industry into livestock feeds. The distilling industry, which uses about 1.6 percent of the nation's grain, most of it corn, now converts almost 100 percent of its grain wastes into feeds. As regards a salable product, the industry is in a fortunate position. The fermentation of grain that results in alcohol also produces, as a by-product, grain residues and liquids that have about three times as much protein as the original grain. Since corn has about 9 percent protein, the nonalcoholic "spent corn stillage," as the industry refers to it, has about 27 percent protein. The levels of many essential amino acids are low, making distillers feeds an unbalanced source of protein, but they play an important role in animal feeding. In 1976, the distilling industry produced 383,500 tons of feeds.

At least one scientist, Dr. Lowell D. Satterlee of the University of Nebraska in Lincoln, thinks these nutritious grain by-products could feed people as well as animals. In a presentation he gave before the Distillers Feed Research Council in 1976, Satterlee described his successful efforts to produce a distillers protein concentrate (DPC). The corn wastes yielded a DPC with 81.06 percent protein, a PER of 1.46, and a fairly high protein digestibility. The wheat DPC had 87.87 percent protein, a 1.66 PER, and higher protein digestibility. Both concentrates were very similar to the starting grain in amino acids and PER. One possible use for wheat DPC is in white bread. In experimental loaves, it did better than soy isolate in improving protein content without decreasing loaf volume. But the ultimate use for DPC and other high-protein additives, Satterlee believes, is in protein blends which, when put in processed foods, will yield all the chemical, nutritional, and functional characteristics desired. To get the right blends, Satterlee feeds data into a computer, which responds with the ingredients. The cookie blend the computer devised includes flour, bean protein concentrate, whey, and corn DPC.

The list of sources of waste protein also includes potato juice from potato starch factories, flour milling by-products, and the partially defatted chopped beef and beef fatty tissue produced in some meat processing operations. Research indicates that each one of these wastes, or a derivative of it, is a good source of protein. Another group of wastes offers comparatively little protein itself but supplies a medium for the growth of high-protein single-celled organisms. A number of vegetable processing wastes fall into this group. The United States is often charged with being a wasteful nation and the volume of protein we presently dump into our waterways or otherwise dispose of indicates the charge is true. But not for long. In the light of new antipollution legislation and the world protein shortage, retrieval of more of our waste protein makes such good sense that the foods of the future are bound to include more items such as whey, blood, and hair. Not under those names, perhaps, but they'll be there. And a waste by any other name tastes just as good—maybe better.

12

PASS THE MUSSEL-STUFFED SQUID, PLEASE

"It was an immense cuttle-fish, being eight yards long. It swam crossways in the direction of the *Nautilus* with great speed, watching us with its enormous staring green eyes . . ."

—*Twenty Thousand Leagues Under the Sea*

THE SQUID, A TEN-ARMED MARINE INVERTEBRATE THAT WAS once regarded as a monster, may soon be a source of protein on American dinner tables, along with a number of other unfamiliar aquatic animals. Not too long ago, these creatures were looked on as "trash" species to be tossed back when they were pulled out of the water along with more popular species. Their lowly status arose from unattractive features such as excess bones, in most cases, although some seem to have been relegated to the trash heap without much reason. But the day of the trash fish has finally arrived. Due to a shortage of popular food fishes, trash fish, now referred to politely as "underutilized species," are being studied by marine scientists for possible use as food. A few have already moved out of the laboratory and onto the fish counter and more will undoubtedly follow.

Besides the squid, some of the underutilized marine and

freshwater species you're likely to find at the fish counter someday soon, if they're not already there, are the skate, the mussel, the red crab, a deepwater clam known as the ocean quahog, the sea urchin, an assortment of small, sardinelike fish called clupeids, and a group of finfish including the pollack, hake, ocean pout, dogfish, Lake Erie white sucker, hogchoker, and croaker. All these creatures exist in abundance in our lakes or off our shores and research indicates all make good eating if properly prepared. A few are already available, although not always under the names given. The Lake Erie white sucker, for instance, which is being test marketed in some areas, is now officially referred to as the freshwater mullet, with the approval of the U.S. Food and Drug Administration. Species with names like the hogchoker and dogfish could obviously benefit from a name change, too.

The ugly squid may be the strongest candidate for increased usage among the underutilized species thanks to a long list of virtues. It's abundant, readily harvestable with existing equipment (most squid, in fact, are caught when fishing boats are trawling for other species), easily cleaned, good-tasting, adaptable to a wide range of traditional recipes, and offers more meat per pound than most other aquatic creatures. At present, though, the squid's faults, while not nearly so specific as its virtues, seem to rule out its wider use. The squid is, well, *funny-looking*. Its tube-shaped body and long tentacles make it resemble a space missile with arms, while its big, human-looking eyes give it a sort of monstrous glare. This peculiar appearance is probably what gave the squid its reputation as a sea monster because it is not really a very dangerous animal. According to Jacques Cousteau, there is no record of a squid killing any human being within the past fifty years. "Lazy and timid," he terms them.

The squid's bad reputation began millennia ago with sea monster stories told by sailors, but it probably got its biggest boost from Jules Verne's classic novel *Twenty Thousand Leagues Under the Sea*. In one episode, the submarine in which the characters are voyaging is attacked by a number of squid, which try to destroy the vessel with their beaks, the small hard structure associated with the mouth. Verne seems to have been weak in biology. Some editions of his book describe the squid as cuttlefish, and illustrations in early editions show an eight-armed octopus. The squid, cuttlefish, and octopus are all cephalopods, members of a quick-moving group of marine mollusks with large eyes and large brains. Intelligence tests indicate that the octopus and the squid are possibly the smartest of all the invertebrates, with a capacity to learn simple mazes and solve rudimentary multiple-choice problems. The only way in which Verne's "squid" resemble real squid is size. Some species of deepwater squid are the world's largest invertebrates, and a dead specimen sixty feet long, including its trailing tentacles, is on exhibit at the American Museum of Natural History in New York City. It's the largest squid ever captured.

No giant squid will turn up on our dinner plates. The squid being suggested as a new protein resource are small species, the length of the body (minus tentacles) ranging up to fourteen inches. Most belong to two related species, *Loligo pealei*, which lives off our East Coast, and *Loligo opalescens*, which lives off our West Coast. Both are believed to be present in large numbers. On four brief research voyages sponsored by the federal government's National Marine Fisheries Service, a single trawler caught 169,000 pounds of squid off the Massachusetts coast. Squid may be even more plentiful off our West Coast, where a small squid fishery, the only one in the United States, already exists. The harvest goes to West Coast restaurants and fish markets or is exported. A major study of the California squid fishery is now underway by California State University's Moss Landing Marine Laboratory. The National Marine Fisheries Service's Northeast Fisheries Center in Gloucester, Massachusetts, has completed a smaller study on the East Coast.

Nutritionally, the squid is an excellent source of protein. The protein content is about the same as that of fish, 16 to 20 percent, but the squid has more usable meat than most other edible creatures. Fish yield about 20 to 50 percent edible flesh, shellfish 20 to 40 percent. Squid, lacking bones (their only rigid internal structure is the small chitinous "pen," which is shaped like an old-fashioned writing pen), have 60 to 80 percent edible flesh. All parts of the body are edible: missile-shaped body, long arms, fins, and even the head if the small hard "beak" is removed. Considering the fact that squid have those large, human-looking eyes, however, you may want to put the head aside and concentrate on the rest of the squid's body. The fat content of the flesh is negligible—from 1 to 5 percent.

Although most Americans disdain squid, the same isn't true in some other areas of the world. "The squid is probably the most universally eaten marine animal except for the United States and Canada," says Robert A. Learson, a food technologist at the Northeast Fisheries Center in Gloucester. According to Learson, squid is eaten in most of the countries bordering the Mediterranean and in many countries in Asia as well. The ugly but nutritious invertebrate is particularly popular in Italy, Greece, and Japan. The word *calamari* that appears on some Italian menus means squid and many Italian restaurants in this country regularly offer it. The foreign fishing vessels that work off our coasts take numerous squid from our waters for use in their own lands. Under the new law setting up a two hundred-mile zone off our coasts over which we exercise fishing rights, squid are one of the few species for which foreign vessels are being permitted to fish. Why? We don't want the squid.

Someday soon, however, our reluctance to eat squid may change. The verdict of taste panels at the Northeast Fisheries Center is that squid is a tasty meat with a firm texture, delicate taste, and a white color that makes it suitable in a range of traditional recipes. In fact, the panel gave the squid such high marks that the Center has decided to make the neglected invertebrate one of its prime candidates for augmenting our fish supply. To that end, they've made up a poster called "The Joy of Cooking Squid" that includes six recipes and a panel showing how to clean squid. If you make all six dishes and so inform the Center, they'll send you a certificate of accomplishment. The poster can be obtained from the Gloucester Center (National Marine Fisheries Center, Marketing Services Division, Dale Avenue, Gloucester, MA 01930). I got a copy of the poster when I visited the Gloucester Center and made one of the recipes myself—squid spaghetti sauce. As Bob Learson had told me, cleaning squid was easy and the spaghetti tasted delicious. Unfortunately, my husband, who is not usually a finicky eater, was so upset by the idea of eating squid (even though he admitted it tasted good) that he made me promise not to fix it again.

I still have two-thirds of a three-pound package of frozen squid sitting in my freezer, but I'm hoping to convince my husband to try it again.

Will most Americans react the same way to squid? It's possible. In a recent survey of over one thousand restaurant owners, the underutilized species these professionals thought their customers would like *least* was the tasty but ugly squid. But I'm betting that the squid's virtues will eventually win it friends, as are scientists at Gloucester and Moss Landing. If you want to experiment with squid, you may be able to find it where I did, in the frozen food case of your supermarket (at, I might add, a very low price). Fish markets in large cities, particularly cities with large Italian, Greek, or Oriental populations, often carry fresh squid. A few canned squid products are usually available in gourmet stores or gourmet sections of supermarkets.

The reason why scientists are looking at squid and other strange aquatic animals with new interest today lies in the shortage of popular food fishes. At the present time, American commercial fishermen catch under five billion pounds of fish per year, about half of which is food fish. When you include the food fish we import and the fish caught by amateur fishermen, we eat about seven billion pounds of fish each year, which works out to over 12 pounds per capita. That doesn't sound like much compared to our per capita

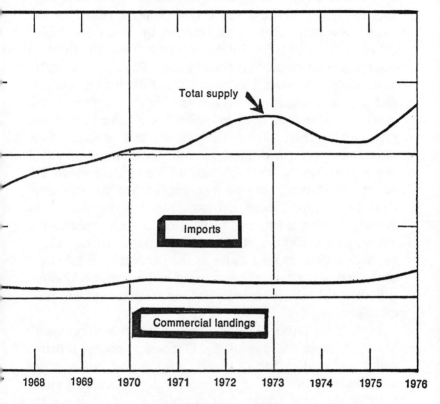

U.S. SUPPLY OF EDIBLE FISHERY
PRODUCTS, 1967-76
(Billion pounds, round weight)

Total supply

Imports

Commercial landings

| 1968 | 1969 | 1970 | 1971 | 1972 | 1973 | 1974 | 1975 | 1976 |

Source: U.S. Dept. of Commerce, Fisheries of the United States, 1976.

beef consumption of 192 pounds (see Chapter 10), but we
have to import about four billion pounds of edible fish each
year to make up the difference between what we catch and
what we eat. Our fish consumption has been rising more or
less steadily since the 1950's, and is expected to continue to
rise, so we'll be importing even more fish in the future, unless,
of course, we can produce more fish ourselves.

The United States has increased its fish catch in the last

decade, but not enough to cover the increasing demand. Most of the other countries where fishing is a big industry have upped their catches much more dramatically. Between 1965 and 1974, for instance, the USSR almost doubled its fish catch. As a result of the bigger catches by most other fishing nations, the world fish catch increased from fifty-three to seventy million metric tons (one hundred fifty billion pounds) in the decade between 1965 and 1975. Fisheries experts predict the world catch could reach one hundred million metric tons of familiar food fishes by 1980. If the United States can get a bigger share of the increase, we will probably meet most of our present demand for fish, but not our projected demand. By 1985, it's predicted, we'll be eating up to 2.3 billion pounds more fish than we eat today. We can, of course, continue to import more and more fish, but the increase in the world catch is expected to level off at some point in the not-too-distant future at around the one-hundred-million-metric-ton level, at least as far as the familiar food fishes are concerned. After that, rising demand and a level world catch will make the fish we eat today far less available and far more expensive.

Luckily for lovers of aquatic protein, the fish story doesn't have to have an unhappy ending. One way to rewrite it that is being suggested by fishery experts is to exploit presently underutilized species such as the ones we now discard. Another facet of the same approach is to utilize more efficiently the fish we already catch. Most of the fish caught in the world today are the popular and familiar food fishes such as flounder and haddock, as well as marine invertebrates, such as shrimp. About thirty species represent 70 percent of the world catch. In large areas of the United States, even fewer species comprise the bulk of the catch. Along our north Atlantic coast, fishermen concentrate on only about nine popular species. The rest of the fish and invertebrates caught by commercial fishing trawlers are simply tossed overboard.

It's estimated that one-half of the catch of American fishing trawlers is thrown overboard after being caught. Many of these fish die after the trauma of capture.

As a result of the emphasis on relatively few fish and invertebrates, some of the most popular species are overfished. Before long, warn marine scientists, they may simply disappear. Along the Atlantic coast of North America, fish in danger of such a fate include haddock, Atlantic halibut, Atlantic herring, mackerel, yellowtail flounder, lobster, Atlantic salmon, sea scallops, and some species of crab. On the Pacific Coast, the arrowhead flounder, sardine, ocean perch, black cod, some salmon, the yellowfin sole, Pacific herring, and Pacific halibut are in a similar plight. Under the new legislation put into effect early in 1977, foreign fishing vessels will be unable to fish for any of these scarce species within two hundred miles of the United States. The foreign vessels, which are much larger and better-equipped than their American counterparts, have often been accused of creating the scarcity of popular food fishes off our coasts. They have undoubtedly added to the problem, but some American fishery experts claim U.S. and foreign fishermen have both contributed to the problem. The new law also mandates quotas for overfished species for the U.S. fishing industry, the quotas to be set by regional councils.

These measures may help bring back most of the scarce food fishes, but another way to insure a future supply of our fishy favorites is to augment them with not-so-beloved fish and invertebrates such as squid. "If we average our fishing efforts over a larger number of species, some of the other species may come back," Bob Learson told me, expressing the view of the National Marine Fisheries Service. "Besides," he added, "we may get to like some of these new species." These arguments are even more popular outside the United States than they are here, particularly in big fish-eating nations like Japan (185 pounds of fish per person per year). Some of their experts claim the world catch could be doubled or even tripled if some of the plentiful underutilized fish and invertebrates were added to it. The leading candidate to increase the world fishery is a tiny—up to two and one-half inches long—invertebrate called the krill that lives in the Antarctic Ocean. So numerous is this small creature that estimates are that one hundred million tons of them could be caught each year.

Some krill are already being caught by the Soviets and the Japanese. The Soviets make it into a reddish, 13 to 20 percent protein substance, Ocean Paste, which is on the market in Russia. It's suggested for paté, fish balls, and salads, among other dishes.

There are rumors that Ocean Paste isn't really very good and that it's piling up in Soviet warehouses, but you can't beat the statistics on krill. From all accounts, it is, as one Japanese scientist put it, "the biggest source of animal protein left in the world today." But putting krill on the dinner table involves problems beyond its acceptability as a reddish paste. Krill is the major food of some endangered animals, for instance, the blue whale, as well as a large part of the diet of some other species, including seals. If man takes vast numbers of krill out of Antarctic waters, what will happen to the nonhuman species that depend on krill? In 1977, the United States began its first intensive study of the tiny invertebrate under the direction of Dr. Mary Alice McWhinnie of De Paul University, a krill expert and veteran Antarctic researcher. The project, which will be headquartered in the Antarctic, will attempt to determine how many krill can be harvested without disturbing the delicate ecosystem of the Antarctic.

Even without krill, though, there are plenty of other underutilized fish in the sea, many within our new two-hundred-mile offshore limit, as well as a number of neglected species in our lakes and rivers. Basically, there are two ways to market these unfamiliar species: You can sell them as whole fish or fillets, or you can convert them into another product. The list of underutilized species being considered for marketing as whole animals is made up largely of invertebrates, led by the versatile squid. Another very likely-looking prospect is the red crab of the Atlantic coast, a one- to three-pound species with delicate, pinkish-white flesh. Once considered uneconomical to catch because of its deepwater habitat, the red crab is now available whole in Maryland fish markets and in a number of East Coast restaurants. A canned, picked meat product is also on the market. The sea urchin, a spiny relative of the starfish, also looks like a winner, although at present most of the urchins captured off our West Coast (urchins also are present off the East Coast) go to Japan. There the orange-colored reproductive organs, which make up one-fifth of the body weight, are considered a delicacy. Sea urchins offer a bonus: their gold-colored roe, which some people claim tastes better than caviar. And then there's the mussel, a neglected relative of the clam and oyster with a narrow dark shell and a good-tasting if rather small mass of flesh. It, too, is available in some fish markets but the best place to find it in the U.S. is a Spanish restaurant, which will offer any number of recipes including mussels. In Spain, the mussel is one of the leading seafoods. The evidence gathered by various fishery researchers indicates that our coasts contain a sufficient number of all these invertebrates to make a sizable commercial fishery possible.

The same is true of two finfish, the ocean pout and pollack, that for obscure and not-so-obscure reasons have been largely ignored by U.S. trawlers off the East Coast as they overfish such species as the flounder. Pout alone account for almost one-half of all the trash fish caught by commercial fishermen off the mid-Atlantic states, according to the University of Rhode Island, which recently began a campaign to push the lowly pout. Pout's firm, white flesh and pleasant taste make it a natural for fresh fillets and it's now being sold in some fish markets in Rhode Island and New York. A brochure furnished with the pout describes how to cook it and gives a few recipes. The pollack is also plentiful off the northeast coast, but until recently it was so little regarded that it hardly paid fishermen to bring it to shore. Now, as a result of a marketing campaign by the Northeast Fisheries Center, it is selling for a reasonable price. A number of fish markets in the East handle it.

Some underutilized species such as the pollack will probably make the big time as a whole animal, but the best use for many neglected fish is their conversion to a more attractive product. An aquatic version of hamburger with far less fat than its meat counterpart, for instance. A machine that has been in use in the fish industry for only a few years is capable of separating flesh and bones and mincing the flesh into a product that can be shaped into various forms. One job it does particularly well is converting the small, bony fishes that make up many of the underutilized species into an item indistinguishable from the flesh of better-known species. Today, the machine is used primarily to make such frozen minced fish items as cakes and sticks from conventional species, but there's no reason to restrict it to these animals. The new machine also removes the remaining flesh from fish after the bulk of it is turned into fillets.

The Northeast Fisheries Center has developed an inexpensive one-pound frozen minced block from underutilized fish that it thinks may make a hit with homemakers looking for a cheap, versatile source of protein. Taste tests and market surveys indicate that the product would do well in the supermarket frozen food case, particularly if a recipe folder is included. The recipes the Center has worked out include chile con carne with two-thirds hamburger and one-third minced fish, fish chowder, and fish spaghetti sauce. The minced fish block goes particularly well with hamburger. In one extensive taste test the Center conducted at a southern university, various proportions of minced fish were mixed with hamburger and served to students in the cafeteria. These veteran hamburger eaters gave the product with 25 percent and 35 percent fish higher ratings for several characteristics than the all-beef patty! The Center has thought up a catchy name for the beef-fish product: beefish.

A number of researchers at other institutions are now experimenting with minced trash fish and the results, in almost all cases, have been enthusiastically received. Dr. Ted S. Y. Koo, chairman of the Department of Fisheries at the University of Maryland's Chesapeake Biological Laboratory, chose a fish with a peculiarly unfortunate name as his minced fish research subject: the hogchoker. A very prevalent species in the Chesapeake Bay, the small (six inches) fish is universally ignored by amateur and commercial fishermen in the area. For Koo's project, a group of his summer students caught a large number of the fish with a trawl net and ran them through a mechanical flesh-bone separator like the one used in industry. The minced flesh that emerged from the machine was formed into fish cakes, which were served at a July 4 picnic for about one hundred fifty people. According to Koo, those who ate the cake came back for "seconds and thirds."

And if they'd been served as hogchoker fish?

No one would have eaten them, believes Koo.

Another fish with good-tasting flesh but an unattractive name is the Lake Erie white sucker, which is small and bony to boot. Abundant in the Great Lakes, it's simply tossed back by most fishermen when it's caught. To make use of the suckers, a group of researchers headed by Dr. R. C. Baker at the State University of New York converted them into minced flesh and then into a fried fish ball and fish chowder. Taste panels liked the dishes and so did a group of professional chefs who sampled them. In one taste test, a panel scored the flavor and tenderness of the sucker fish balls higher than fish balls made from the popular, expensive, and overfished flounder. Fish balls are a nutritious product, having over 16 percent protein. The chowder, which is largely liquid, has about 6 percent protein. Both the balls and chowder are due to be test marketed soon, with the name of the sucker, as indicated earlier, appearing as freshwater mullet. A one-pound minced block of sucker flesh similar to the product developed at Gloucester is already being test marketed in Rochester, New York.

Mincing underutilized species isn't the only way to get people to eat this neglected protein resource. In Japan, a machine-made form of fish cake called kamoboko that resembles a fishy version of some of our luncheon meats is highly popular. It's estimated that about 25 percent of the total fish catch in Japan is converted into these cakes, for which a wide variety of fish are suitable. Although kamoboko contains some additives, principally starch, it has more protein and less fat than most of our meat sausages and wieners, which are loaded with nonprotein ingredients, including fat. Could kamoboko catch on over here? Judging from the kamoboko I bought in a Japanese market in New York City, the Japanese version is a bit too bland and rubbery for our tastes (the rubbery texture is considered a plus in Japan), but a product more to our tastes could easily be devised. A kamoboko machine is in operation at the Gloucester laboratory and an American-style fish sausage may someday be developed there.

Still another way to get underutilized fish into our diet is via fish meal. About one-half of the United States catch and about one-third of the total world catch is converted into fish meal, but little of it goes to people. The recipients are animals, principally chickens in this country. Until 1972, the major fish used for fish meal was a species of anchovy caught off Peru, but the Peruvian fishery has fallen off dramatically in recent years and much of the fish meal we use today in this country comes from a small species native to our coasts, the menhaden. Fish meal has about 60 to 65 percent protein and a good balance of amino acids, and some experts are beginning to question its use solely for animals. "We feel that fish for animal feed is not an efficient use of a good protein source," says Dr. John Spinelli of the National Marine Fisheries Service's Pacific Utilization Research Center in Seattle. He and a team of scientists at the Center recently made a high-quality protein powder designed for human use from Pacific hake, an underutilized species. A white, odorless product, it can be added to a number of foods, including frankfurters, whipped cream topping, angel food cake, and brownies. In none of these items, claims Spinelli, can the marine source be detected.

The United States doesn't produce enough fish meal even for its animals today, but some underutilized fish off our coasts may be sources of fish meal and high-protein fish powder. One group of fish that look like good candidates for these products is the clupeids, small, sardinelike fish that are plentiful in the Gulf of Mexico, which also is the site of a major menhaden fishery. A survey of the clupeids by Dr. Edward D. Houde of the Rosenstiel School of Marine and Atmospheric Sciences at the University of Miami indicates that some of them could support commercial fisheries. Among the most numerous species are the thread herring, the round herring, the Spanish sardine, and the anchovy. None is presently being taken in the Gulf for commercial purposes, although there is a commercial anchovy fishery off California.

There will always be a fishing industry, but there's more than one way to harvest a fish. In the future, many marine biologists and other experts believe, farm-raised fish will play a much bigger role in satisfying our taste for aquatic protein. Today, fish farming, or aquaculture, supplies only about 1.4 percent of the fish Americans eat, and most of those fish are concentrated in a few species. Have you eaten a trout in a restaurant lately? Almost all the trout sold commercially today come from trout farms, most of them in the western United States. About half the catfish eaten in this country are raised on farms, a similar proportion of crawfish, and about 40 percent of the oysters. A small percentage of salmon are also provided by fish farms. As far as aquaculture goes in this country, though, that's about it. All the other fish we eat are wild species caught in oceans, lakes, and rivers.

In many countries, fish farms play a much bigger role in providing protein. Worldwide, about six million tons of aquatic species, including seaweed, are produced on farms, which adds up to a little less than one-tenth of the world fish catch. About 66 percent of that is finfish, 16 percent mollusks, such as the mussel, the rest seaweed and crustaceans, primarily shrimp. Three-fourths of the world's aquaculture is concentrated in countries in the Indo-Pacific region, where aquaculture is an old, old story. Japan, the leader in world aquaculture, raises a wide range of aquatic species, including marine shrimp, oysters, various marine finfish, and seaweed.

Japan's aquaculture practices are the most sophisticated in the world and it raises some species which have never been successfully cultured elsewhere. Sophistication isn't really necessary for success in aquaculture, however; the milkfish, a herbivorous tropical species, has been reared for centuries in ponds by farmers in Southeast Asia. A low-priced fish, it furnishes a significant protein resource for the population in areas in which it's raised.

Some of the aquatic species being successfully farmed throughout the world may play a bigger role in putting protein on our own table in the future, along with species that have proved themselves here. In a 1976 report prepared by the United States Department of Commerce, the department which includes the National Marine Fisheries Service, increased production of all marine creatures now being farmed here was recommended, along with the cultivation of some old ocean favorites such as the lobster. In addition, though, the report suggested farming some animals that are little known here: freshwater prawns, a relative of the popular marine shrimp, and herbivorous fishes such as the milkfish. The Commerce Department doesn't just recommend the cultivation of such species. Under the Office of Sea Grant, it gives money to private and public groups which are researching the cultivation of new and old species of marine animals.

Most of that money is being spent on tried and true animals like the oyster and the shrimp, but in recent years, a new species has swum into Sea Grant funding: the freshwater giant prawn, which is said to taste like a cross between a shrimp and a lobster. In 1976, prawn research at the South Carolina Marine Resources Center in Charleston received a $360,000 Sea Grant, which was augmented by $197,000 in funds from the State of South Carolina and private industry. Five other research groups also received Sea Grants to raise giant prawns. The reason behind this official interest in the giant prawn is the Great American Shrimp Shortage. The marine shrimp is the number one species we catch in terms of value, but we can't catch enough to satisfy our needs. We import more shrimp than we catch, for a total of about five hundred million pounds per year. And our future demand for shrimp is expected to go up, up, up. The Japanese, who like shrimp even more than we do, manage to fill their shrimp demand by rearing them, but shrimp farming is so difficult that no commercial firm in this country has succeeded in making a profit out of it, although some are still plugging away in Florida. A partial solution, some experts now believe, is the giant prawn, which is much easier to farm than the marine shrimp.

The giant prawn, a freshwater relative of the marine shrimp, is found in rivers and lakes all over the world, including the United States, but the species selected as an aquaculture candidate is *Machrobrachium rosenbergii*, a native of Malaysia. In the 1950's, Chinese-born American scientist Dr. Shao-wen Ling, who was then with the United Nations Food and Agricultural Organization (FAO), came across the prawn in Southeast Asia, where it has been farmed and eaten for many years. A creature about six inches long, *M. rosenbergii* looks something like a big version of the familiar marine shrimp, except for its long legs. It bears its young in salt water, but spends most of its life in fresh or brackish water. Ling began studying the animal and eventually succeeded in breeding and rearing it in captivity, a necessary first step in most aquaculture projects. Thus far, marine shrimp have never been bred in captivity. In 1966, Takaji Fujimura, a biologist with the Hawaii Department of Land and Natural Resources, heard about Ling's work and asked him for some specimens. Ling sent thirty-six. Six survived and became the foundation of a new fish-farming industry in Hawaii that now includes a half dozen small commercial farms and a state-operated hatchery with millions of young *M. rosenbergii*. Fujimura still directs the hatchery, as well as an extensive research project involving all facets of *M. rosenbergii* culture.

Giant prawns are available in Hawaiian fish markets and a number of the better Hawaiian restaurants. One Honolulu restaurant, Chez Michael, offers them as a Friday night special. They're not cheap, though, retailing for almost five dollars a pound, a price that puts them somewhere between shrimp and lobster.

Still, shrimp and lobster continue to find eager buyers even at high prices, and the expectation is that *M. rosenbergii* will do the same if and when it comes on the market in the continental United States. Considering the half dozen or more successful research projects involving the giant prawn that are now in existence, as well as a commercial enterprise located in Puerto Rico that is already selling prawns to restaurants, the giant prawn may be almost here. To get a firsthand look at the creature we'll probably be eating someday, I visited one of the most successful research projects, the South Carolina Marine Resources and Research Commission's giant prawn hatchery in Charleston. The success of this particular project has exceeded expectations, since *M. rosenbergii* is a tropical species and South Carolina, while indisputably southern, is not tropical. The secret of the warmth-loving prawn's adaptability to these somewhat northerly latitudes is an indoor winter hatchery in which the young prawns are taken to the postlarval or juvenile stage before being turned out into outdoor ponds in the spring. Once they've reached this size, they grow even more rapidly in South Carolina ponds than they do in Hawaii. The total outdoor growing season doesn't extend much beyond six months, but that's enough to bring the prawns to eating size—about twenty to a pound—within a single year. As Drs. Paul Sandifer and Theodore Smith, the directors of the project, see *M. rosenbergii* farming in South Carolina, the prawns will be reared to perhaps an inch in length in an indoor hatchery, possibly supported by the state. Then they will be transferred to farmers, who will grow them to eating size in outdoor freshwater ponds, using chicken feed as food.

"What we're proposing is another farming operation, like raising catfish," Sandifer told me. "It won't be more difficult than raising catfish, except for the temperature constraints."

The hatchery Sandifer and Smith have designed incorporates a number of new ideas, at least as far as *M. rosenbergii* is concerned. One of them is a place to hide. *M. rosenbergii*, like its shrimp relatives, is an aggressive animal given to cannibalism at close quarters. In many prawn farms, the mortality rate reaches 50 percent, or even higher. Sandifer experimented with horizontal strips of window screening in some prawn tanks to give the creatures a place to avoid their fellow prawns and a possible death by cannibalism. Mortality rates dropped significantly in the tanks with screening and it is now used in most of the tanks. "With the screening, we can put in three to four times as many animals as we did at first and still get good survival," said Sandifer as we stood looking at a tank with a number of minute, transparent postlarvae. When he motioned toward one of them, it sprang away like a mosquito.

The water for all the prawns, said Sandifer, is recycled and heated to eighty degrees. For the larvae he uses a commercial product called Instant Ocean familiar to saltwater aquarium hobbyists. "I haven't used a drop of real seawater in three years." The older breeding stock live in freshwater. "Here's some mating behavior," he said, pausing beside a big tank containing a number of adults. One large prawn hovered over a small one. "That's the male," said Sandifer, indicating the large prawn. "He's protecting the small one, the female, because she's about to molt. If he didn't do that, the others would eat her. As soon as she molts, they'll mate and then, in about a day, she'll extrude an egg mass. The male transfers a sperm packet to the female. When her eggs pass over the sperm, they're fertilized."

I was impressed with the hatchery, but, as I told Sandifer, I had expected to see something larger in the way of prawns because of the name giant prawn.

Eating prawns, he explained, are harvested at a small size, "but they get bigger if you keep them long enough. I have one over here I call my 'gee whiz' specimen." We walked over to a small separate tank that held what looked like two impressive-looking prawns, both the size of skinny lobsters. "She weighs about a half pound and is about a foot long," noted Sandifer.

"That thing with her is her molt." He also showed me an even larger dead specimen on display which belongs to another species, *M. carcinus*. Native to Florida, it reaches three pounds and is sometimes caught for food. Why not raise *carcinus* for food? "They have a nasty disposition. Also, the larvae are smaller and harder to raise. They grow more slowly, too." Eventually, though, some of the other members of the Sandifer-Smith team expect to cross *M. rosenbergii* with other native species to obtain a hybrid that is better adapted to colder temperatures. If they succeed, the giant prawn will join the very few aquatic animals that can be considered truly domesticated.

Another farm-bred freshwater creature that might make a good substitute for some of the marine shrimp in our diet is the crawfish, a shrimp-sized crustacean native to the United States. Louisiana farmers have been rearing the red swamp crawfish (*Procambarus clarki*) and the white river crawfish (*P. blandingi acutus*) in ponds for a quarter-century but there was little interest in either eating or rearing crawfish in other areas until recently. But in the last few years, pilot crawfish farms have been started in Texas, Mississippi, Arkansas, and Missouri, and people outside Louisiana are beginning to realize what residents of that state have known for years: Crawfish are one of the best-tasting animals that ever came out of the water. I almost missed eating my first crawfish when I visited Baton Rouge, because I arrived during a chilly February at the end of the coldest winter in memory in the South. Since the crawfish harvest begins when the weather turns warm in January or February, there were no Louisiana crawfish available. Luckily, some crawfish from Florida were imported to fill the gap and I ordered a plate of them at Ralph and Kacoo's Restaurant. The crawfish arrived in a towering pile that I finished to the last morsel. A substitute for shrimp? Crawfish are *better* than shrimp.

The latest innovation in crawfish farming is polyculture, the farming of two or more species in the same area. Polyculture has been practiced in Asia for millennia, but it is a new idea here. At Louisiana State University in Baton Rouge, Dr. James Avault, one of the nation's few experts on crawfish, is growing the crustacean in the same ponds with three finfish: catfish, buffalo fish, and paddlefish. The crawfish and buffalo fish, both bottom-dwellers, prowl along the floor of the pond, eating natural pond organisms and leftover catfish rations that drop out of the floating baskets in which the cats are kept. The paddlefish live off plankton in the upper part of the water. Yields of the polyculture project, which is being operated under a Sea Grant, are extraordinary. At best, a fish farmer could expect to harvest about two thousand pounds of fish or crawfish per acre if just one species was raised, but the four-species ponds at Louisiana State University yield more than forty-seven hundred pounds of crawfish and fish per acre. Avault believes any fish farmer could duplicate his success with the crawfish and compatible finfish in a climate similar to that of Louisiana.

Polyculture may help the crawfish find a larger number of fans in the United States but our freshwater crustacean will probably never be as popular as the marine shrimp in spite of the crawfish's delectable flavor. The crawfish has less flesh per animal than the marine shrimp and cannot be processed by machine, two factors which would make it expensive to market on a nationwide basis. In Louisiana much of the crawfish harvest, which is evenly divided between wild and farm-reared animals, is sold live within the state's borders at a fairly inexpensive price. Still, efforts are being made to develop a crawfish processing machine and it may someday be feasible to sell these animals at a reasonable price outside a few southern states. A fairly sizable market for the prawns may exist in areas where they are completely unfamiliar today. In a marketing test conducted in Houston, Texas, in 1975 by Dr. James C. Carroll of the University of Southwestern Louisiana in Lafayette, seafood markets and restaurants sold a total of 46,521 pounds of crawfish in a four-month period, even though crawfish had not been available in Houston before. Reaction to the new food item was termed "very enthusiastic."

In the northernmost part of the United States, a fish-farming project involving another neglected native invertebrate is just getting underway. The animal is *Mytilus edulis*, the mussel, and if there ever was an unexploited species, this is it. Mussels are common off our eastern and western shores, but few are harvested due to an inexplicable ignorance on the part of the American public concerning the delights of mussels. Recently, several groups, including the National Marine Fisheries Service and the Maine Department of Marine Resources, have begun advocating a mussel fishery and the mollusk is now available in some fish markets along the East Coast. Mussels, according to these groups, are delicate and nutlike, neither as fishy as fish nor as clammy as clams. Meanwhile, however, a team composed of researchers from the universities of Maine and New Hampshire and the Maine Department of Marine Resources, together with Edward A. Myers, a resident of Maine who recently started the nation's first mussel farm, have pooled their efforts under a Sea Grant to develop mussel farming in the Gulf of Maine.

Mussell farming is new here, but mussel farms in Spain are the most successful form of aquaculture in the world. Yields from these Spanish farms reach an almost incredible five hundred thousand pounds of mussels per acre per year, which is by far the largest yield of any aquaculture enterprise. Taking advantage of the mussel's tendency to cling to submerged objects, the Spanish farmers suspend ropes from rafts in protected bays. Baby mussels attach themselves to the ropes in a continuous line and grow to adulthood. No supplementary food is needed for the crop, which feeds on organisms available in the water. When the mussels are eating size, they are harvested simply by pulling up the rope and removing the mussels. A mechanized adaptation of the rope-raft system is being developed by the United States researchers for use in the Gulf of Maine.

It takes more than a mechanized system to sell a new product, though, and the group also has prepared dozens of recipes for mussels. As they put it, "Try them, you'll like them." You could probably say the same for most of the other under-utilized species. In fact, some of the new sources of aquatic protein are so good, we may end up liking them better than our old overfished standbys.

13

EATING THE UNTHINKABLE

> "I have been assured by a very knowing
> American of my acquaintance in Lon-
> don, that a young healthy child, well
> nursed, is, at a year old, a most deli-
> cious, nourishing and wholesome food,
> whether stewed, roasted, baked or
> boiled; and I make no doubt that it
> will equally serve in a fricassee or a
> ragout."
>
> —JONATHAN SWIFT

ONE OF THE FAVORITE RECIPES OF TWENTY-SEVEN-YEAR-OLD
home economist Carol Miller and her husband, Dennis, is
termite pilaf. Yes, termite pilaf, as in termite, the small, wood-
eating insect pest. The recipe, which Mrs. Miller devised her-
self as part of a senior project at California State University
in San Luis Obispo, includes rice, sesame seeds, onion, beef
bouillon and, instead of meat, one cup of termites. Other
insect-based dishes Mrs. Miller dreamed up for her project
include Bee Won Ton (a taste panel liked this one the best),
which incorporates a quarter-cup of bees, and Jiminy Bread,
a bread with a crunchy texture derived from a cup of roasted
grasshoppers. The 'hoppers are a close relative of the cricket,
hence the name Jiminy in honor of the famous cricket. The
Millers eat these and other insect-based dishes several times
a week.

Even uncooked insects taste good, insists Carol Miller, who likes to serve freeze-dried bee larvae as croutons in salads. "Delightfully crisp," she says.

Most of the recipes she has worked out revolve around creatures like the termite, the bee, and the grasshopper because they are readily available and relatively bland in taste. In a number of conventional recipes, the presence of insects is scarcely discernible. Then why use them? Many insects have a high protein content, Miller points out in her paper "Introducing Insects into the American Diet," making them an important nutritional resource. Dried grasshoppers, for instance, have some 60 percent protein, compared to about 43 percent for dried beef round. True, all the grasshopper protein cannot be utilized by the human body, but insects have supplied protein to many cultures from the era when man first emerged from the apes to the present day. Apes also eat insects, as do monkeys. Dr. Jane Goodall, the British animal behaviorist, watched wild chimpanzees in Africa poking sticks into termite nests and withdrawing the stick full of insects, which they ate with apparent enjoyment. Carol Miller's paper discusses entomophagy, the practice of eating insects, among American Indians, including the Modoc, Pitt River, and Diggers.

"In fact," she says, "insects might well be considered America's original 'Soul Food.'"

The Millers are not the only present-day Americans who have returned to the nutritious practice of entomophagy. Californians tend to be particularly enthusiastic about eating insects, perhaps because residents of that state are a little more prone to try something new. In 1975, a book called *Butterflies in My Stomach* by Dr. Ronald L. Taylor, a Los Angeles County pathologist, was published by Woodbridge Press, a California firm. As the book relates, butterflies literally have been in Taylor's stomach, along with an amazing variety of other insects, their larvae and eggs. His favorite? The greater wax moth, which he calls "far tastier and more nutritious than snack items such as potato chips." Taylor reports that

WHOLE-BODY ANALYSES OF VARIOUS
INSECT SPECIES (DRY-WEIGHT BASIS)

ORDER, GENUS, OR SPECIES	PROXIMATE COMPOSITION (%)	
	PROTEIN	FAT
House fly pupae		
Musca domestica	63.1	15.5
Termites		
Isoptera	45.6	36.2
Adult locusts		
Melanoplus	75.3	7.2
Oxya	67.8	4.5
Oxya	74.7	5.7
Schistocerca gregaria	61.8	17.0
Schistocerca paranensis	51.1	18.4
Nomadacris septemfasciata	63.5	14.1
Sphenarium	50.6	

Source: DeFoliart, *Bulletin of the Entomological Society of America*, 1975.

the protein content of insects ranges from a modest 8.0 percent in the larvae of some butterflies (about the same as the protein content of rice) to an impressive 63.4 percent in roasted spiders (actually an arachnid, not an insect). Insects high in protein include housefly pupae (the dormant stage of an insect), 63.1 per cent; dried grasshoppers, 60.0 percent; fried termites, 45.0 percent; and smoked caterpillars, 38.1 percent. Compare these figures with those for cooked beef, 19.7 percent protein, and cooked pork, 29.4 percent, and you can see why people like Taylor and Miller see insects as an alternative source of protein in a protein-short world. Another factor in insects' desirability as food is their abundance. When numbers alone are considered, insects are more plentiful than any other group of animals, excluding some microscopic creatures, and insects are far and away the most numerous in terms of species. Insects, as Taylor puts it, have "provocative and promising possibilities for a world whose food supply cannot keep pace with its exploding population."

Some cultures, as indicated earlier, already have discovered insect food. Taylor delved exhaustively into the subject of insect eating in various times and places and discovered that practically every large group of insects has been eaten by at least one culture. The most popular, far and away, is the nutritious and widely available grasshopper, which has been enjoyed by everyone from eighth-century-B.C. Assyrians to nineteenth-century American Indians. Today, grasshoppers (a group which includes the huge swarms of migratory locusts) are still eaten by tribesmen in Africa, Australia, South America, and New Guinea, as well as residents of some areas of Japan. In Japan, however, grasshopper eating is dying because insecticides have vanquished most of the grasshoppers. The number two world favorite is the termite, which is a dietary item in South America, India, Indonesia, Thailand, Africa, and Australia, among other areas.

Not satisfied with this literary survey of entomophagy, Taylor collected some insects and fed them to students in his classes at the University of Southern California, where he is an assistant professor. The students liked the live pupae of Taylor's own favorite insect, the greater wax moth, and the fried larvae of the flour beetle, *Tribolium*, which reminded some of them of pork cracklings and roasted sunflower seed. To test the more widely available canned insects, Taylor set up a taste panel. The most popular products proved to be fried agave worms (the larvae of the Skipper butterfly of Mexico), fried grasshoppers ("good on pizza," suggested one panelist), and just about any chocolate-covered insect, including grasshoppers, ants, and caterpillars. The least popular were fried silkworms, fried butterflies, and roasted caterpillars.

Here's Taylor's recipe for fried *Tribolium*:

Fry fresh insects for a few minutes in a shallow layer of hot vegetable oil. Drain on paper towels. Add a little salt and serve hot.

If the thought of fried *Tribolium* or fried grasshoppers makes you feel a little jumpy, Taylor brings up an interesting point: Americans eat a fair amount of insects during their lives, a certain level of insect contamination being unavoidable. Almost all fresh lettuce contains aphids, and weevils and beetles commonly infest flour and rice. "Anyone who has consumed much rice has eaten the rice weevil," states Taylor calmly. As for prepared foods, allowable infestation is quite high in some items. For instance, the Food and Drug Administration allows five insects or insect parts per one hundred grams of apple butter, fifty insect fragments per one hundred grams of peanut butter. As one who has eaten many a gram of peanut butter, I must have eaten my share of insects, too. Not to mention the occasional whole cockroach that turned up in dishes I prepared in my former apartment in New York City.

So if a cockroach or ant accidentally falls in your stew or picnic lemonade, don't worry. Taylor claims such insect additions literally *enrich* food.

If you are willing to try putting insects in your food, you might want to know that Taylor and Barbara J. Carter, a California high school science teacher, are coauthors of a new Woodbridge Press book, *Entertaining with Insects*, which they say is the only recipe book for insect cookery. It contains a number of intriguing dishes ranging from hors d'oeuvres to dessert. If you make these recipes, you don't have to be afraid that your family or guests will be startled by what looks like a bug that wandered onto their plate by mistake; like Carol Miller's recipes, the Taylor-Carter dishes make the insect ingredient almost unrecognizable. Why not just serve the dish and tell everyone afterward? Another helpful feature of the book is that it is built around insects that are readily available, principally mealworms (the larvae of the *Tenebrio molitor* beetle), crickets, and bees. If you're not a keeper of small reptiles, as I am, you may not be aware that the first two insects are available from some pet stores and, for a lesser price, from a number of mealworm and cricket "farms" throughout the country, many of which the book lists. Once you have a supply of these creatures, it's possible to raise more yourself in a basement or even a corner of your apartment with the proper equipment.

Bees can be kept by the amateur too, of course, although you'll need more room and chutzpah to do it. In many communities, live bees can be obtained from commercial beekeepers, who are often listed in the yellow pages of the telephone book.

Another resident of California who suggests entomophagy to help fill the world protein gap is Dr. Roy Snelling, an entomologist at the Los Angeles Museum of Natural History. Dr. Snelling regularly fries up a mess of grasshoppers or termites and consumes them with enjoyment. Insects, he believes, are a food of the future for Americans, even if beefsteak is still around. "I'm quite serious," he told me in a recent interview. "Insect tissues are a very high source of protein, much higher than beef or any of the other proteins we normally eat." Snelling has eaten many of the insects found in southern California, including cockroaches, crickets, termites, and grasshoppers.

"I don't care much for cockroaches," he said. "They have a peculiar taste. So do crickets. But freshly roasted grasshoppers are quite edible. I collect them myself and keep them alive for three or four days and feed them on corn meal. That takes out any bitter flavor they may have from eating alfalfa or other grasses. It's sort of like corn-fed beef. To cook them, I put a little oil in a skillet, brown the insect up good, put a cover on the skillet and let them simmer for twenty minutes or so. They're absolutely delicious." Termites—the three-quarter-inch ones found in the western states—also taste good fried in this manner, Snelling said.

A humane way of killing the insects before frying them, he suggested, is to drop them in boiling water for a few seconds.

But aren't insects—cockroaches, in particular—rather unclean? I asked. After all, cockroaches eat garbage.

"We eat hogs and *they* eat garbage," countered Snelling. Actually, he said, most insects, including the cockroach, are very clean, an opinion with which other entomologists concur. Insects, these experts point out, are constantly cleaning themselves, an activity you can provoke by dropping a little water on any handy insect. More likely than not, it will try to clean off the offending drops, a procedure it follows with dirt and other foreign substances. Also, the diet of many insects, including the grasshopper, consists of plants, which makes them particularly attractive from the point of view of cleanliness.

Snelling had other encouraging words for the neophyte insect-eater. "It's not much different from eating crustaceans like shrimp or lobster," he said. "Crustaceans aren't so far removed from insects by a long shot. Next time you eat a shrimp, just think that you're eating a relative of the insect." Entomologist Snelling is correct. Marine crustaceans such as the shrimp, lobster, and crab belong to the large phylum of invertebrates called the arthropods, to which the insects also belong. This makes insects more closely related to shrimp than, for instance, to earthworms, which belong to an entirely different phylum of invertebrates. Both crustaceans and insects have a hard covering, six legs, and a three-part body. The next time you see a lobster creeping about its tank in a restaurant, see if you don't think it looks rather like an overgrown insect.

If small insects are ever eaten with anything like the enthusiasm we reserve for big crustaceans, Snelling believes insects will have to be cultured so that they are available in sufficient quantities. "In other words, you'd have to farm them," he says. "I'm thinking particularly of termites. Some of the large species we have out here in the West could actually be cultivated on recycled waste paper. They do very well on that. After all, we already culture things like crickets and earthworms for fish bait. A lot of the technology exists right now; it's just a matter of, first of all, someone sitting down and doing it, and in the second place, creating a market."

One entomologist actually *has* farmed an insect with an eye to eventual mass marketing as human food. He is Dr. Edwin W. King of Clemson University in Clemson, South Carolina; the creature he has cultured is the face fly. He rejected Snelling's choice, the termite, as too troublesome to raise, an objection he also makes to everyone's favorite insect, the grasshopper. The face fly is a little larger than the common housefly and unlike its relative, it prefers dry pasture and plains. It breeds in cow dung. This rather unpleasant habit was one of the qualities that focused King's attention on the face fly, since manure provides a cheap and plentiful food on which to rear the animal in the laboratory. Other advantages are rapid growth, relatively large size, and immunity to disease. In addition, its protein content puts it in the high range for an insect: about 52 to 53 percent.

After four years of work on the project, King has come up with some intriguing figures, which he provided me. Theoretically, he claims, a face fly colony grown on the cattle dung available from an acre of cattle in a feedlot can increase the protein production of the feedlot operation from about 436,680 pounds of protein for the beef alone to over 800,000 pounds of total protein: beef plus face flies. The production rate for the fast-growing fly in King's mini-production plant —"three rooms in a broken-down old building," as he puts it—was one gram of pupae per square inch of manure surface per day. King estimates that a factory consisting of fifty-eight buildings on an acre of land could produce 885,490 pounds of pupae per year, or 398,500 pounds of protein. An acre of beef cattle, he points out, produces only about 566,280 pounds of protein per year, which dresses out (minus inedible parts) to the figure of 436,680 pounds of protein given earlier.

But as King himself says, the ultimate test of any food is not figures but what happens when somebody eats it. Thus far, face fly pupae have been eaten experimentally by two hundred catfish, twenty chickens, and King himself. It seemed to harm none of them and both the catfish and chickens thrived on a mixed diet of pupae and commercial meal. The chicks, in fact, did better on the mixed diet—soybean meal and pupae—than they did on their regular soybean meal diet. As for King, he finds pupae taste like sawdust. But then so do a lot of processed food ingredients before flavoring is added. "When I'm putting on a show in public, I mix the pupae with cinnamon and then it tastes like cinnamon," he says.

As King's experiments indicate, insects may make an acceptable diet for livestock, even if we can't bring ourselves to swallow the little creatures. A group at the University of Wisconsin has achieved good results by feeding fly pupae —a quiescent stage in the life of some insects—to baby chicks. When the pupae formed the major protein source in the chicks' diet over a seven-week experiment, the chicks gained almost as much weight as chicks on a conventional diet. The pupae, which have from 61 to 63 percent protein, were grown on a substance that is very common on farms: manure. Harvested by a flotation process, the manure-fed pupae were dried and then ground to a "meal" for incorporation into the chick diet.

A couple of other creepy, crawly creatures that are not insects but are often linked with them in the popular mind might also serve as good alternative sources of protein. One such creature is the spider, an arachnid. Arachnids have eight legs, insects six, among other differences. In a recent issue of *Smithsonian Research Reports*, a publication put out by the Smithsonian Institution in Washington, D.C., I came across a brief article describing the use of giant wood spiders (*Nephila maculata*) as a protein source in New Guinea. The article was based on the work of Drs. Michael and Barbara Robinson, who had tried the giant spiders for themselves and found them quite tasty—"like peanut butter without the objectionable consistency," as they put it. Their favorite spider recipe is a traditional New Guinea one. First you collect fat female spiders in the open end of a hollow green bamboo stick, the other end of which is stopped with a folded leaf. Then place the stick with both ends stopped in the embers of a fire for ten to fifteen minutes, or until the green bamboo is blackened. When you take the roasted spiders out, their hard skins have split and they are ready to be eaten, with or without legs.

I wrote the Robinsons, who are now at the Smithsonian's Tropical Research Institute in the Canal Zone studying, among other creatures, a spider relative of *Nephila maculata*, and Michael Robinson wrote back, enclosing a paper in which they described the species. The giant wood spider lives up to its name, being almost two inches long. Some of its eight legs are over three inches long. Egg-carrying females may weigh three grams (.11 of an ounce) or more, a hefty weight for a spider. "We've also eaten, in New Guinea, the larvae of Cerambycids," Michael Robinson wrote. "The beetles lay their eggs in dead trees and the grubs grow to a large size feeding on the wood. Just before they pupate, when they are about three to four inches long, the grubs are full of stored fat. They are collected and sold in the markets like bundles of firewood. We fried them in butter after first removing the strong mandibles and squeezing out the gut contents. They were very tasty, but I must admit that I had some bad dreams about maggots gnawing away at my brain."

Giant wood spiders like the ones the Robinsons ate are creatures of the tropics, but other large spiders might serve as a protein source in North America. A more likely food prospect among our native invertebrates, however, is the earthworm, which is enjoying something of a boom in the United States, according to *National Geographic* magazine. In one recent year, the worm raising industry—and it *is* an industry—produced an estimated one billion worms. At least one earthworm farmer, Ronald Gaddis, president of North

American Bait Farms in Ontario, California, the nation's biggest commercial worm operation, has become a millionaire within the past decade. The new popularity of earthworms is partly due to their suitability for fish bait, but also to the increased interest in organic gardening. Put a passel of worms in your garden and they aerate the dirt and manufacture fertilizer, claim gardeners, obviating the need for chemical fertilizers.

What does all this have to do with protein? Well, there's a reason why fish gulp down earthworms with such gusto. According to a study conducted in 1976 by California State Polytechnic University for North American Bait Farms, the red earthworm sold by the California firm has 11.8 percent protein on a wet-weight basis, the way they are in their natural state, and 58.9 percent protein on a dry-weight basis. To produce dried worms, the researchers baked them in a conventional oven at three hundred degrees for thirty minutes. Beef round dried in the same manner had only 43.1 percent protein. The California Polytechnic researchers also learned that worms are eaten today by tribes in South Africa, New Guinea, and Australia. Added confirmation of the popularity of earthworms among Australian tribesmen comes from *National Geographic*, which says some Australian species of worms grow to several feet in length and an inch or so in breadth. When baked, they are said to taste like pork sausage. And look like sausage too, no doubt.

We don't have any giant earthworms in the United States, but the red 2½-inch-long earthworm sold by North American Bait Farms makes an even better recipe component, according to that firm. The bigger nightcrawler (it gets to five inches or so) they dismiss as too "wormy" in taste. The red worm, on the other hand, is essentially flavorless, giving it an advantage in traditional recipes. To prove the point, North American sponsored the first red worm cooking contest in 1976, dubbing the competition Ver de Terre Recipe Contest. *Ver de terre* means "earthworm" in French. The first prize of five hundred dollars went to a not particularly French-sounding dish called Applesauce Surprise Cake, which included one cup of chopped, dried earthworms. Other winning entries included an earthworm omelet and earthworm-stuffed peppers. Another contest was being planned for 1977 with the same top prize money. Someone soon, according to Jerome LaTour, North American's public relations director, a recipe book with all the winning recipes will appear. By the way, LaTour says all the personnel at North American Bait Farms have eaten earthworms in various recipes. The consensus of opinion: Earthworms don't have much taste.

If you're planning to whip up a few worm recipes of your own before the recipe book appears, North American Bait Farms offers this advice on worm preparation: "Worms must first be thoroughly washed in cold water and then boiled to remove stray bits of soil and to kill any undesirable bacteria." After that, you're on your own.

Difficult as it is to envision the average American family dining on Applesauce Surprise Cake or Bee Won Ton, these dishes stand a better chance of appearing on the table than some of the other protein sources being suggested in various quarters. Dogs, for instance. To a visitor from outer space, eating pets undoubtedly has much to recommend it. Meat tends to have much the same protein content and balance of amino acids no matter what the source; many cultures have enjoyed dog cuisine; and there are certainly enough surplus dogs around to furnish meat. In fact, we have so many extra dogs that each year we kill an estimated one million in shelters alone. Some end up in pet food. Well, why not eat surplus dogs ourselves? A scientist asked that question recently in the letters column of *Science*, the journal published by the American Association for the Advancement of Science. Here's what he said:

> Food shortages are a current reality in many areas of the earth, and the dietary component in shortest supply is protein. Simultaneously, as Feldmann reminds us, we are confronted with a major dog problem: too many dogs are uncontrolled,

unwanted, and unowned. The measures proposed to control the dog problem—leash laws, population control, and public education—are expensive and ineffective. But with a more enlightened viewpoint, could we not consider excess dogs a significant nutritional resource that deserves our attention?

Western man seems to suffer a total mental block about the concept of eating dogs. However, in many cultures, dogs have been a traditional component of man's diet. In much of Oceania, dogs have been preferred over pork. Early British visitors to Hawaii and Tahiti described Polynesian methods of dressing and cooking dogs and compared the product favorably with English lamb.

Undoubtedly some will raise objections because, in Western culture, dogs have been sanctified as pets. Such objections are without merit. In the first place, as Feldmann and Beck point out, stray and unowned dogs (which are not pets) are a major part of the dog problem. Second, many animals (chickens, ducks, rabbits, calves) fill dual roles as pets and as food. In my experience such pets have been every bit as delicious as their relatives with whom I have had no personal relationship. Third, could anyone bestow a higher honor on a pet than to make it part of oneself?

—DONALD B. MILLER
Corvallis, Oregon

This all sounds logical enough, but I thought I detected a resemblance to satirist Jonathan Swift's "A Modest Proposal," the eighteenth-century essay in which Swift advocated the marketing of Irish children as a food item to alleviate their parents' poverty. Swift, of course, wasn't serious and I didn't think Miller was either. But when I said so in an article, Miller wrote me saying: "I was, and am, perfectly serious." At any rate, the best rebuttal to the Miller position came in a letter from another scientist, M. Ian Phillips, which was printed in a subsequent issue of *Science*. No, on checking, I see it wasn't from Phillips but from his dog (this letter definitely wasn't serious). It ran in part: "He [Miller] says that several cultures eat dog. On this argument I could add that several cultures have also been known to eat man. Let the unthinkable remain uneatable."

Another rather unthinkable beast being suggested as a protein source is the guinea pig, that cute little rodent that children keep as a pet in this country. In a symposium held at the Rockefeller Foundation in 1975, Dr. James McGinnis of Washington State University in Pullman, advocated both the rabbit and the guinea pig as alternative protein sources because they can utilize some cellulose, the fibrous plant material man cannot digest. Guinea pig meat may sound revolting to us but it has a long history as a food item. When the Spaniards entered Peru in the early sixteenth century, they found the Incas keeping captive guinea pigs, which are native to Central and South America, as a meat source. The little rodents are still used as food in some areas of South America. *Pan American World Health,* a journal published by the World Health Organization, recently reported that an estimated seven million guinea pigs are harvested annually for food in Peru. That number is slated to go up, as efforts are currently being made to increase the supply of guinea pig meat. The journal put the protein content of guinea pig flesh at 19 percent, about the same as that for beef, pork, and poultry.

Rodents are popular as food all over South America, which has a unique population of large rodents found nowhere else in the world. If you've eaten a dish in South America and didn't recognize the meat, it may well have been one of these giant rodents. Among the rodents eaten in various areas are the twenty-pound agouti, a long-legged animal with a solid brown coat; the paca, which looks much like the agouti but is larger and has a coat with spots and stripes, and the pacarana, a stocky, large-headed rodent weighing up to thirty pounds with a spotted and striped coat. The first two are common in forested areas, but the pacarana, which is found in the Andes, is now rare, at least partly because of the popularity of its flesh. All these rodents are hunted as wild game, but the largest rodent in the world, the one-hundred pound capybara, is not only a game animal but is now managed as a semidomesticated food resource on ranches in Venezuela. Capybara flesh has special appeal in Venezuela because a salted, dried product is a traditional Lenten dish that can be eaten even when other meats are prohibited. The illogical logic behind this is that the capybara (*Hydrochoerus hydrochaeris*), a semiaquatic species that spends much of its time in the water, is not really meat but something more akin to fish.

Because of the ease with which the plant-eating rodent can be raised on ranches with cattle, some experts believe the capybara has a big future as a food item in South America and, possibly, as an export item for other countries, including the United States. In 1972, two South American scientists, Juhani Ojasti and Gonzalo Padilla, told the North American Wildlife and Natural Resources Conference in Washington, D.C., that the capybara deserved a closer look as a food source. Already, they pointed out, capybaras were making more money for some Venezuelan ranchers than cattle because the rodents produce more food than cattle for the amount of vegetation they consume. On one ranch, a total of fifteen thousand capybaras were harvested in one recent year. And there were twenty-eight thousand left! Some advantages capybara have as ranch animals are their peaceable dispositions, slow movements, and herding instincts. Capybaras, in fact, act much like miniature cows, except for their fondness for swimming.

Their flesh doesn't taste like beef, though. "Indifferent," said Charles Darwin when he tasted capybara meat in 1832 on his voyage around South America, and that seems to sum up the majority opinion. Ojasti and Padilla suggest turning capybara meat into smoked sausages or a canned meat product to widen its appeal.

Canned capybara will probably make it into the supermarket before grasshoppers or worms. We already eat a few wild rodents, such as the squirrel, but excluding the American Indians, we have never eaten insects or worms. Not only that, but most of us regard these invertebrates with something like disgust as a food item, no matter how nutritious, cheap, and available they might be. Overcoming this cultural barrier might be too big a job even for Madison Avenue. One nutritionist who advances this viewpoint is Dr. Sohan Manocha of Emory University in Atlanta. He advocates the wider use of insect species to fill the protein gap in his 1975 book, *Nutrition and Our Overpopulated Planet*, but not for the United States. "Man is not a metabolic engine," he told me. "We do not consume nutrients, we consume diets. Food is part of people's culture. In places where insects are an accepted part of the diet, they should be cultivated more for food. Why not farms where locusts could multiply and you could harvest them like meat?"

He added firmly, "But insects will never be accepted in the United States."

Needless to say, the new breed of entomophagists doesn't agree although they see the cultural barrier as their most formidable obstacle. Some have plans to overcome it. Carol Miller's way of combating what she sees as an "irrational prejudice" is to concoct appetizing recipes in which common insects can be used as a protein supplement, rather than as a major substitute for meats. For her project, she picked insects that not only are readily available but have a bland taste, making it easy to hide their taste with other ingredients. Bees, for instance, she describes as having a "bland straw- or hay-like flavor." Bee larvae, too, are bland, so much so they can even be eaten raw. Termites are another bland insect, with the added advantage that they are small and rather unobtrusive in shape. "If you use wild rice in my termite pilaf recipe, you can't tell the termites from the shafts of rice," she claims. Grasshoppers, fairly big insects, are harder to hide, but their taste is innocuous enough that ground-up grasshoppers are undiscernible in Carol Miller's Jiminy Bread. Miller thinks bland-tasting, ground or otherwise disguised insects could be used to increase the protein content of a range of prepared foods such as soups and grain products like pasta and cereal.

There is still another way for insects and worms to creep quietly into the American diet, though: the cocktail hour. Americans, it seems, will eat almost anything with cocktails, their qualms washed away by alcohol. Most of the canned and bottled insects now on the market are believed to be used for cocktail parties. But if you want some good cocktail party insects, don't rely on the prepared ones available in gourmet food stores, which are almost universally scorned by entomophagy enthusiasts. Make your own insect hors d'oeuvres from wild or cultivated insects or worms. There are a number of intriguing insect hors d'oeuvre recipes in the Taylor-Carter book, including Cricket Crisps and Salted Garlic Mealworms. I think I'm going to try some of these for a cocktail party of my own, but I wish someone else would make them. I don't mind eating insects or worms, but preparing something that is liable to crawl or jump right out of the kitchen really is creepy.

14

A CHANGE IN DIET?

"Free 72-oz. Steak if this dinner is
eaten in one hour."

—Advertisement for
Big Texas Steak Ranch,
Amarillo, Texas

Is a change in diet ahead for the American consumer?
High prices for traditional sources of protein and the possibility of reducing malnutrition in developing countries are powerful arguments for turning to a wider selection of protein
sources. Many of these "new" protein sources—actually, most
are an old story in some areas of the world—are so appealing
we'll probably learn to like them as well as we do meat and
dairy products. But there is another argument for a change in
the American diet and it may turn out to be the most powerful
one in encouraging us to include some of the nontraditional
protein sources in our meals. In recent decades, a number of
scientific studies have uncovered a sinister link between the
American diet and the American way of death. Or, to be more
exact, between the high consumption of fat, particularly animal fat, and sugar on the part of populations of North America and Western Europe and the major killer diseases in those
areas. The fat- and sugar-rich "affluent diet," as some scientists refer to it, is a goal to strive for in the eyes of most of the
world but it may be the principal reason for early death in
developed countries.

"In living the better life we are, ironically, exposing ourselves to greater health risks," says Dr. Beverly Winikoff, of the Rockefeller Foundation.

In our nation, as in most of Western Europe, the big three among killer diseases are coronary heart disease, cancer, and stroke. Coronary heart disease leads the list in our country. A general term for diseases involving the coronary arteries through which the heart supplies itself with blood, coronary heart disease often involves a "heart attack" in which the supply of blood to the heart is cut off. Although the incidence of coronary heart disease is dropping in the United States, it still takes some seven hundred thousand lives every year. Cancer is the second leading cause of death, stroke the third in the United States. Stroke results from an impaired supply of blood to parts of the brain and also involves the arteries. In both coronary heart disease and stroke, the mechanism that cuts off the supply of blood to the heart or brain is the deposition of fat on the inner wall of the arteries, a condition called atherosclerosis or, more popularly, hardening of the arteries. Included in the fatty deposits is a fatlike substance called cholesterol which is naturally present in the blood and other organs but which is also acquired from foods of animal origin.

Recent studies indicate that in young American males, the deposition of cholesterol in the arteries begins as early as the age of two. By their twenties, what one U.S. researcher calls "massive infiltration" of the arteries by cholesterol is underway.

An indication of the extent to which cholesterol is being deposited in the arteries and, by inference, of the risk of coronary heart disease and stroke, is the blood or "serum" level of cholesterol. This can be determined today by a simple test. Dr. Ancel Keyes, a U.S. researcher, published an article in 1956 in the medical journal *Journal of Chronic Diseases* which showed that countries whose populations had blood cholesterol levels averaging 220 milligrams and above had a high incidence of coronary heart disease. Countries whose populations had blood cholesterol levels averaging 200 milligrams and below had little coronary heart disease. In Yugoslavia, for instance, the average blood cholesterol level was 160 and there was a low incidence of coronary heart disease. The United States, with an average blood cholesterol level of 240, was among the nations with the highest incidence of coronary heart disease. East Finland, which had an average blood cholesterol level of 265, had the world's highest incidence of coronary heart disease. The best-known population study associating coronary heart disease with blood cholesterol levels is the Framingham Study, a twelve-year-long investigation of five thousand subjects carried out in the Massachusetts town of Framingham under the auspices of the federal government's National Heart and Lung Institute. It indicated that up to age fifty-five, the risk of developing coronary heart disease leaps upward with increasing blood cholesterol levels.

Most of the research attention on the link between heart disease and the presence of fat in the blood has been focused on cholesterol but another type of fat is now believed to be associated with heart disease: triglycerides. Triglyceride, which is simply the scientific name for fat, is the predominant fat in both our diet and our own body tissues. Research now indicates that elevated blood levels of *both* cholesterol and triglycerides are linked with heart disease.

What does all this have to do with diet?

Another group of population studies, some of them dating back almost to the turn of the century, links a certain type of diet with a high level of blood cholesterol and triglycerides. Again, most of the work has been done with cholesterol. One of the first studies, and still one of the most dramatic, was made by a Dutch scientist living in Indonesia. He noted low blood cholesterol levels among the Indonesians and Japanese, both of whom received most of their protein from vegetable sources and fish. When he checked Japanese who worked on Dutch passenger liners where the passengers and crew ate diets rich in meat and other animal products, however, he found a rise in blood cholesterol level. In fact, the levels were as high as those of the Dutch passengers. Almost a half century later, a number of researchers studying various populations showed that diets containing the type of fat present in animal products result in raised cholesterol levels, while diets rich in vegetable fats result in lowered cholesterol levels.

The same link between animal fat and cholesterol has been demonstrated with laboratory animals. In one typical study, adult rabbits fed a diet with about twice the level of cholesterol (all of it from animal products) as that in the average American diet developed cholesterol deposits in their arteries in just two months. After five months, the arteries were covered with a solid plaque of cholesterol.

Some of the most provocative studies involving cholesterol have been carried out with populations that have moved from an area where a low-cholesterol, low-fat diet is popular to an

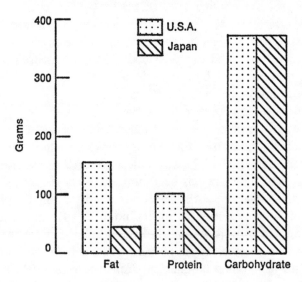

COMPARISON OF PER CAPITA CONSUMPTION
OF NUTRIENTS IN THE U.S.A. AND JAPAN

Source: National Nutritional Survey, Japan, 1969; National Food Situation, U.S.A., 1968.

area where a high-cholesterol, high-fat diet is in vogue. In one study of this nature, Japanese who moved from Japan, where fat consumption is about one-third of that in the United States, to either Hawaii or California had a higher incidence of coronary heart disease than the Japanese population in Japan. The rate of increase was moderate in Hawaii but high in California, where the diet of the new immigrants underwent the biggest change. Further evidence of the effect of the affluent diet is the fact that coronary heart disease incidence in Japan, although still low, has tripled over the past fifteen years, a period that corresponds to a rise in meat eating. Another study of a population that has moved from one area to another shows that Jews from Yemen, where there is a low incidence of coronary heart disease, approach the high incidence among the Israelis after the Yemenite Jews have spent twenty years in Israel.

Fat appears to be the major factor in the Western diet that predisposes people to a higher level of blood cholesterol and triglycerides and, eventually, to an increased risk of heart disease and stroke, although sugar probably plays an accessory role. The word "fat" usually conjures up a picture of the solid white material clinging to beef and pork, or the yellowish lumps beneath the skin of poultry. Both those substances are fat, but most of the fat we eat is less easy to see and eliminate. We acquire most of it in our quest for protein. Fat is scattered through some cuts of beef, particularly those from juicy, grain-fed carcasses. In general, the more expensive the cut of beef, the higher the fat content, with cuts like T-bone steak and standing rib roast containing much more fat than round steak or heel of round. But the beef product with the highest amount of fat is, as you might expect, cheap hamburger. Veal (young beef) has markedly less fat than beef. Excluding a few very fatty cuts, pork and lamb have about the same fat content as beef. The fattest meat cut is bacon; frankfurters and most lunch meats are also high in fat. Among poultry, duck and goose are high in fat, chicken and turkey low. In addition to fat, most flesh meats contain moderately high levels of cholesterol. Organ meats and blood are very high in cholesterol. Protein concentrates made from organs and blood, however, have little fat or cholesterol.

Fat is high in most dairy products, including milk, cheese, butter, ice cream, and eggs, and in several vegetable sources, principally nuts. Most dairy foods also have moderate levels of cholesterol. The only dairy products with low fat and cholesterol levels are nonfat dry milk and whey, and the products made from them, and cottage cheese.

Our fat consumption has risen significantly since the early part of the century. Today we eat about 125 pounds of fat per person per year, which works out to about 156 grams (5.4 ounces) per day. In 1910, we ate about 100 pounds of fat per person. The rise in dietary fat amounts to about two and one-half tablespoons a day per person and is accounted for chiefly by the increase in meat consumption, followed by an increase in the use of salad oil and cooking oil. In all, about 42 percent of the calories in our diet now come from fat, an increase of about 10 percent since 1910. By way of contrast, developing countries acquire less than a fourth of their calories from fat. Some countries, however, exceed even our fat consumption, including Canada, Denmark, and New Zealand.

Our sugar consumption has risen even more dramatically. We now eat about 100 pounds per person each year, compared to about 76 pounds in the first decade of the century. The present world average is 44 pounds per person.

Americans are obviously consuming large amounts of fat and sugar, but as far as the relationship between fat and atherosclerosis is concerned, there is evidence that it isn't the *amount* of fat we eat, so much as the *kind* of fat we eat, that is important. Two kinds of dietary fat have been implicated in high blood cholesterol and triglyceride levels: cholesterol and the so-called saturated fats, most of which are solid fats of animal origin. The American diet, unfortunately, is rich in both. According to Dr. Donald B. Zilversmit, professor of nutritional sciences and biochemistry at Cornell University, we eat from 600 to 700 milligrams of cholesterol every day if

FAT CONTENT AND MAJOR FATTY ACID COMPOSITION OF SELECTED FOODS (IN DECREASING ORDER OF TOTAL SATURATED FATTY ACID CONTENT WITHIN EACH GROUP OF SIMILAR FOODS)

[IN PERCENT]

Food		Fatty acids		
	Total fat	Total saturated	Total monoun-saturated	Total polyun-saturated
Animal fats:				
Chicken	100.0	32.5	45.4	17.6
Lard	100.0	39.6	44.3	11.8
Beef tallow	100.0	48.2	42.3	4.2
Avocado	15.0	2.0	9.0	2.0
Beef products:				
T-bone steak (cooked, broiled—56 percent lean, 44 percent fat)	43.2	18.0	21.1	1.6
Chuck, 5th rib (cooked or braised—69 percent lean, 31 percent fat)	36.7	15.3	17.5	1.5
Brisket (cooked, braised, or pot roasted—69 percent lean; 31 percent fat)	34.8	14.6	16.7	1.4
Wedge and round-bone sirloin steak (cooked or broiled—66 percent lean; 34 percent fat)	32.0	13.3	15.6	1.2
Rump (cooked or roasted—75 percent lean; 25 percent fat)	27.3	11.4	13.1	1.2
Round steak (cooked or broiled--82 percent lean; 18 percent fat)	14.9	6.3	6.9	.7
Cereals and grains:				
Wheat germ	10.9	1.9	1.6	6.6
Oats (puffed, without added ingredients)	5.5	1.0	1.9	2.2
Oats (puffed, with added nutrients, sugar covered)	3.4	.6	1.2	1.4
Barley (whole grain)	2.8	.5	.3	1.3
Domestic buckwheat (dark flour)	2.5	.5	.8	.9
Cornmeal, white or yellow (whole-ground, unbolted)	3.9	.5	.9	2.0
Shredded wheat breakfast cereal	2.5	.4	.4	1.3
Wheat (whole grain, Hard Red Spring)	2.7	.4	.3	1.3
Wheat flakes breakfast cereal	2.4	.4	.3	1.2
Rye (whole grain)	2.2	.3	.2	1.1
Wheat meal breakfast cereal	1.4	.3	.1	.7
Wheat flour, all purpose	1.4	.2	.1	.6
Rice (cooked brown)	.8	.2	.2	.3
Bulgur from Hard Red Winter wheat	1.5	.2	.2	.7
Oatmeal or rolled oats, cooked	1.0	.2	.4	.4
Rye flour	1.4	.2	.1	.6
Cornstarch	.6	.1	.1	.3
Rice (cooked white)	.2	.1	.1	.1
Farina (enriched, regular, cooked)	.2			.1
Corn grits, cooked	.1			.1
Dairy products:				
Nondairy coffee whitener (powder)	35.6	32.6	1.0	
Cream cheese	33.8	21.2	9.4	1.2
Cheddar cheese	32.8	20.2	9.8	.9
Light whipping cream	32.4	20.2	9.6	.9
Muenster cheese	29.8	19.0	8.7	.7
American pasteurized cheese	28.9	18.0	8.5	1.0
Swiss cheese	27.6	17.6	7.7	1.0
Mozzarella cheese	19.4	11.8	5.9	.7
Ricotta cheese (from whole milk)	14.6	9.3	4.1	.4
Vanilla ice cream	12.3	7.7	3.6	.5
Half and half cream	11.7	7.3	3.4	.4
Chocolate chip ice cream	11.0	6.3	2.6	.4
Canned condensed milk (sweetened)	8.7	5.5	2.4	.3
Ice cream sandwich	8.2	4.7	2.6	.5
Cottage cheese (creamed)	4.0	2.6	1.1	.1
Yogurt (from whole milk)	3.4	2.2	.9	.1
Cottage cheese (uncreamed)	.4	.2	.1	
Eggs:				
Fried in margarine	15.9	4.2	7.2	1.9
Scrambled in margarine	12.6	3.7	5.5	1.4
Fresh or frozen	11.3	3.4	4.5	1.4
Fish:				
Eel, American	18.3	4.0	9.0	2.7
Herring, Atlantic	16.4	2.9	9.2	2.4
Mackerel, Atlantic	9.8	2.4	3.6	2.4
Tuna, albacore (canned, light)	6.8	2.3	1.7	1.8
Tuna, albacore (white meat)	8.0	2.1	2.1	3.0
Salmon, sockeye	8.9	1.8	1.5	4.7
Salmon, Atlantic	5.8	1.8	2.7	.5
Carp	6.2	1.3	2.7	1.4
Rainbow trout (United States)	4.5	1.0	1.5	1.4
Striped bass	2.1	.5	.6	.7
Ocean perch	2.5	.4	1.0	.7
Red snapper	1.2	.2	.2	.4
Tuna, skipjack (canned, light)	.8	.2	.2	.2
Halibut, Atlantic	1.1	.2	.2	.4
Cod, Atlantic	.7	.1	.1	.3
Haddock	.7	.1	.1	.2

FAT CONTENT AND MAJOR FATTY ACID COMPOSITION OF SELECTED FOODS (IN DECREASING ORDER OF TOTAL SATURATED FATTY ACID CONTENT WITHIN EACH GROUP OF SIMILAR FOODS—(Continued)

[IN PERCENT]

Food	Total fat	Fatty acids Total saturated	Total monoun-saturated	Total polyun-saturated
Fowl:				
Chicken (broiler/fryer, cooked or roasted dark meat)	9.7	2.7	3.2	2.4
Turkey (cooked or roasted dark meat)	5.3	1.6	1.4	1.5
Chicken (broiler/fryer, cooked or roasted light meat)	3.5	1.0	.9	.9
Turkey (cooked or roasted light meat)	2.6	.7	.6	.7
Lamb and veal:				
Shoulder of lamb (cooked or roasted, 74 percent lean; 26 percent fat)	26.9	12.6	11.0	1.6
Leg of lamb (cooked or roasted, 83 percent lean; 17 percent fat)	21.2	9.6	8.5	1.2
Veal foreshank (cooked or stewed, 86 percent lean; 14 percent fat)	10.4	4.4	4.2	.7
Nuts:				
Coconut	35.5	31.2	2.2	.7
Brazil nut	68.2	17.4	22.5	25.4
Peanut butter	52.0	10.0	24.0	15.0
Peanut	49.7	9.4	22.9	15.0
Cashew	45.6	9.2	26.4	7.4
Walnut, English	63.4	6.9	9.9	41.8
Pecan	71.4	6.1	43.1	17.9
Walnut, black	59.6	5.1	10.8	40.8
Almond	53.9	4.3	36.8	10.1
Pork products:				
Bacon	49.0	18.1	22.8	5.4
Sausage, cooked	32.5	11.7	15.1	3.9
Deviled ham, canned	32.3	11.3	15.2	3.5
Liverwurst, braunschweiger, liver sausage	32.5	11.0	15.5	4.1
Bologna	27.5	10.6	13.3	2.1
Pork loin (cooked or roasted, 82 percent lean; 18 percent fat)	28.1	9.8	13.1	3.1
Ham (cooked or roasted, 84 percent lean; 16 percent fat)	22.1	7.8	10.4	2.4
Fresh ham (cooked or roasted, 82 percent lean; 18 percent fat)	20.2	7.1	9.5	2.2
Canadian bacon (cooked and drained)	17.5	5.9	7.9	1.8
Chopped ham luncheon meat	17.4	5.7	8.3	2.2
Canned ham	11.3	4.0	5.3	1.2
Salad and cooking oils:				
Coconut	100.0	86.0	6.0	2.0
Palm	100.0	47.9	38.4	9.3
Cottonseed	100.0	26.1	18.9	50.7
Peanut	100.0	17.0	47.0	31.0
Sesame	100.0	15.2	40.0	40.5
Soybean, hydrogenated	100.0	15.0	23.1	57.6
Olive	100.0	14.2	72.5	9.0
Corn	100.0	12.7	24.7	58.2
Sunflower	100.0	10.2	20.9	63.8
Safflower	100.0	9.4	12.5	73.8
Shellfish:				
Eastern oyster	2.1	.5	.2	.6
Pacific oyster	2.3	.5	.4	.9
Ark shell clam	1.5	.4	.3	.3
Blue crab	1.6	.3	.3	.6
Alaska king crab	1.6	.2	.3	.6
Shrimp	1.2	.2	.2	.5
Scallop	.9	.1		.4
Soups:				
Cream of mushroom (diluted with equal parts of water)	3.9	1.1	.7	.8
Cream of celery (diluted with equal parts of water)	2.3	.6	.5	1.0
Beef with vegetables (diluted with equal parts of water)	.8	.3	.3	
Chicken noodle (diluted with equal parts of water)	1.0	.3	.4	.2
Minestrone (diluted with equal parts of water)	1.1	.2	.3	.5
Vegetable (diluted with equal parts of water)	.9	.2	.3	.4
Clam chowder, Manhattan style (diluted with equal parts of water)	.9	.2	.2	.5
Table spreads:				
Butter	80.1	49.8	23.1	3.0
Margarine (hydrogenated soybean oil, stick)	80.1	14.9	46.5	14.4
Margarine (corn oil, tub)	80.3	14.2	30.4	31.9
Margarine (corn oil, stick)	80.0	14.0	38.7	23.3
Margarine (safflower oil, tub)	81.7	13.4	16.1	48.4
Vegetable fats (household shortening)	100.0	25.0	44.0	26.0

Source: *Dietary Goals for the United States,* 1977.

we consume typical North American fare. Not all that cholesterol later appears in our blood, but work Zilversmit has carried out indicates that we probably absorb about 300 milligrams of it per day. Since the body itself synthesizes some 600 to 1,000 milligrams of cholesterol per day, we may boost the total cholesterol metabolism in the body by as much as 50 percent through diet.

Our consumption of saturated fats is estimated at about 38 percent of all the fats in our diet, or 47.5 pounds per year. Experiments indicate that some saturated fats help raise blood cholesterol and triglyceride levels. Polyunsaturated fats and monounsaturated fats, the two other major classes of dietary fat, have not been associated with raised blood fat levels; some experiments indicate, in fact, that the polyunsaturates actually lower blood fat levels. Recent research also indicates that sucrose—ordinary sugar—raises triglyceride levels.

The links between dietary fat and sugar, on the one hand, and blood fat levels and atherosclerosis, on the other, have been known for decades, but research carried out within the past few years now implicates dietary fat in some common forms of cancer in this country, too. Dr. Ernest L. Wynder, president of the American Health Foundation, a private research organization with offices in New York City and Valhalla, New York, believes our daily diet probably plays a much larger role in cancer than food additives, substances which are often condemned as cancer-causing agents. The new evidence on which Wynder and a number of other scientists base this belief associates a high level of dietary fat with an increased incidence of colon (large bowel) cancer and breast cancer. Colon cancer is the second leading cause of cancer death among U.S. men, breast cancer the leading cause of cancer death among U.S. women. Studies show that in countries where the population eats a diet with a high fat content, the incidence of both colon and breast cancer soars. Conversely, colon and breast cancer incidence drop in areas where comparatively little fat is eaten. The graphs for geographic incidence of the two diseases are almost identical, with the same

Western nations and nations with Western culture, such as New Zealand, clustered near the top, the same Asian and South American nations grouped at the bottom.

At or near the top are Canada, the United States, New Zealand, Australia, Denmark, and the Netherlands, all among the leaders in meat and dairy food consumption. At the bottom are Thailand, Japan, the Philippines, Colombia, and Ceylon, all with diets based primarily on vegetable protein and/or fish. Most of the nations with a low incidence of colon and breast cancer are developing countries with a low level of industrialization, but Japan is heavily industrialized. Japan's inclusion in the group of countries with low incidence of breast and colon cancer indicates that some factor other than those associated with industrialization must be responsible for the difference in death rates from breast and colon cancer. Genetic predisposition? No, because studies show the incidence of colon and breast cancer rises in Japanese who emigrate to the United States and adopt a Western diet. Also, groups such as the Seventh-Day Adventists in the United States who practice vegetarianism have less breast cancer and colon cancer than the rest of the U.S. population.

An environmental factor must be at work, and the only one researchers have been able to single out is diet. Americans eat three times as much fat as the Japanese and our fat intake, unlike theirs, is mainly of animal origin. One population study indicates that beef consumption is linked with cancer but others indicate that the total level of fat, both saturated and unsaturated, contributes to elevated levels of colon and breast cancer in Western nations, or nations where Western diets prevail.

Animal studies confirm the findings of population studies. A number of investigators have shown that when laboratory animals are put on a diet high in both saturated and unsaturated fats, they have an increased number of breast and colon cancers.

How can fatty food lead to cancers of two such different

organs as the colon and the breast? Scientists don't have the answers as yet, but there are a number of plausible theories. The American Health Foundation proposes that colon cancer arises when fats alter the composition of bacteria in the digestive system, promoting the production of cancer-causing agents. Specifically, the normal anaerobic and aerobic bacteria become more anaerobic, enabling them to better degrade bile salts and cholesterol. Bile acids, one product produced by the degradation process, are thought to promote tumor formation. Eventually, the theory says, these events lead to the production of cancer-causing agents, or carcinogens. The theory has some solid evidence behind it, much of it accumulated by researchers at the American Health Foundation. Populations that have high colon cancer death rates have been shown to excrete high levels of cholesterol metabolites and bile acids. In one study, colon cancer patients excreted more bile acid and bacteria than a control group. Several other controlled studies show that human subjects given a high-meat, high-fat diet secrete much higher levels of fecal bile acids, cholesterol metabolites, and bacteria than subjects given a low-meat, low-fat diet.

The American Health Foundation theory with regard to breast cancer is that a high-fat diet elevates blood levels of a female hormone called prolactin, which has been shown, in earlier work, to promote mammary cancer. The crucial aspect of the prolactin elevation, however, is not simply its level but its relationship with another female hormone, estrogen. When levels of the two hormones reach a certain critical threshold relative to each other, the theory goes, tumors that are already present as tiny changes in breast tissue are stimulated into growth. One of the American Health Foundation studies on which this theory is based shows that an antiprolactin drug abolishes the tumor-promoting effect of a high-fat diet but that an antiestrogen drug does not have the same effect. Another study shows that blood levels of prolactin increase in laboratory animals on high-fat diets.

Colon and breast cancers show the clearest link with fat

consumption but some of the more recent research also implicates the affluent diet in a high incidence of cancers of the womb and pancreas. Lesser ailments such as benign enlargement of the prostate are also believed to be associated with a fatty diet.

Since we acquire most of our fat as we ingest protein, over-consumption of protein itself may actually play a role in cancer. In a study carried out in India, groups of rats were fed either a high-protein or low-protein diet, with all the protein coming from milk casein. Each group of rats was also given a daily dose of aflatoxins, a potent liver carcinogen. Although the rats on the low-protein diet had a higher death rate than those on the high-protein diet, none developed liver tumors or even precancerous signs of liver tumor. One had a kidney tumor. But 50 percent of the rats on the high-protein diet had liver tumors or other types of tumors, and all the rest had precancerous liver conditions. Recently, Dr. T. Colin Campbell, a professor of nutritional biochemistry at Cornell University, and two colleagues, Rachel S. Preston and Johnnie R. Hayes, found that the reason why the low-protein diet had this effect on rats is that the aflatoxins were much less likely to bind to nuclear macromolecules within the cells, a process thought to be a step in the production of cancer.

Based on his work and that of others, Campbell has devised a theory to explain how overconsumption of protein may produce cancer. From 70 to 90 percent of cancers, he pointed out in a recent article in *Human Ecology Forum*, a publication put out by Cornell, are now believed to be produced by chemicals and environmental factors. The generally accepted theory is that after ingestion, the chemicals undergo metabolism by enzymes before they become carcinogens. During metabolism, they bind tightly to DNA, the part of the cell nucleus that carries genetic information. Protein, speculates Campbell, first increases the enzyme activity that converts the chemicals to carcinogens, and then promotes the binding of the carcinogens to the DNA. His own research shows that the enzyme activity required to produce a carcinogen from aflatoxins was greatly reduced on a low-protein diet.

"Most certainly," he writes, "the low protein levels used in our experimental diets are unreasonably low. Are there, though, dietary levels of protein between the unreasonably low and those commonly high levels that could also reduce tumor formation?"

Although a causal relationship between the high level of fat or protein in our affluent diet and the leading killer diseases isn't established as yet, many nutritionists and other scientists involved with nutrition believe that diet presents a clear "risk factor." Their advice: Change the affluent diet *now*. If we wait, they argue, it will be too late to save many American lives. This message was emphasized by the U.S. Senate's Select Committee on Nutrition and Human Needs, which published a report on its findings concerning the American diet in early 1977. Our diet, claims the report, *Dietary Goals for the United States*, constitutes "as great a threat to the public health as smoking." "The risks associated with eating this diet are demonstrably large," notes Dr. D. M. Hegsted, professor of nutrition at Harvard University's School of Public Health, one of the scientists who helped prepare the report. "The question to be asked, therefore, is not why should we change our diet, but why not?"

The Worldwatch Institute, a private research organization with offices in Washington, D.C., takes the same stand. "A comprehensive review of dietary impacts on health reveals a persuasive case for dietary change," asserts "Worldwatch Paper 9," published in December, 1976.

What changes do these experts recommend?

Less meat, fewer dairy products, less fat of all kinds, less sugar, less salt (a risk factor in heart disease). More vegetable protein sources, more low-fat animal protein sources such as veal, chicken, and fish, more starchy carbohydrates, substitution of some unsaturated fats for saturated fats. Specific recom-

DIETARY GOALS FOR THE UNITED STATES AS SET BY THE SENATE SELECT COMMITTEE ON NUTRITION AND HUMAN NEEDS

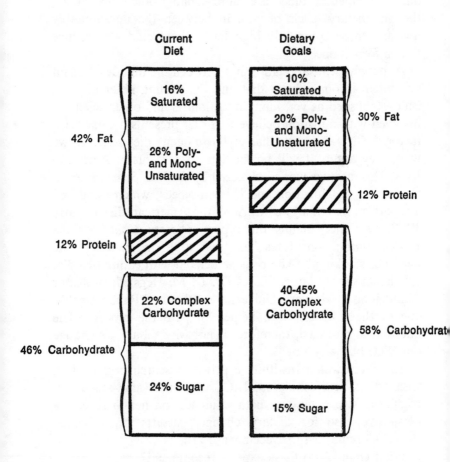

Source: *Dietary Goals for the United States,* 1977.

mendations include most of the alternative sources of protein or their by-products, particularly grass-fed beef (less fatty than grain-fed beef), low-cholesterol meat and cheese analogs, vegetable protein-carbohydrate sources such as grains, beans, seeds, and roots, and highly polyunsaturated oils, for instance, safflower and sunflower oils. One group of foods that emerges with new luster is the starchy carbohydrates, those old no-no's of a thousand slimming diets. Now it appears the starchy carbohydrates have been unjustly maligned. New evidence indicates that starch not only does not make you fat but that a diet rich in starchy grains, seeds, roots, and beans (many of which, of course, are also rich in protein) can also make you healthier than a diet rich in animal fats.

"Most population groups with a low incidence of coronary heart disease consume from 65 to 85 percent of their total energy in the form of carbohydrate derived from whole grains (cereals) and tubers (potatoes)," notes the Senate Committee on Nutrition and Human Needs, quoting Drs. William E. and Sonja J. Connor in *Present Knowledge in Nutrition* (Nutrition Foundation).

The new regard for starches even encompasses pasta, which I had always vaguely assumed was a "bad" food, even though it's one of my favorites. But a study made by Italian researcher Mario Mancini and described in *Dietary Goals for the United States* indicates that southern Italian workingmen, who eat little animal or dairy fat but a high level of starchy foods, including pasta, have lower blood cholesterol and triglyceride levels than their countrymen in northern Italy or Americans, Swedes, Britons, or Swiss, all of whom eat much higher levels of saturated animal and dairy fats. "High-carbohydrate diets are quite appropriate for both normal individuals and for most of those with hyperlipidemia (high levels of fat in the blood), provided that the carbohydrate is largely derived from grains and tubers, that an energy excess is not consumed and that adiposity does not result," notes *Dietary Goals*. "The use of high-carbohydrate diets by civilized man has an historical basis, is economically sound and shows clear indications of causing less, rather than more, disease." In another area of the publication, the consumption of beans, peas, and lentils, along with a number of other foods (fish, shellfish, bread, skim milk, uncreamed cottage cheese, and breakfast cereals) that obtain less than 20 percent of their calories from fat is specifically recommended.

And remember, such a diet is recommended for our own affluent country, not simply for developing nations.

Actually, the "new" diet that nutritionists are recommending is an old diet, even in this country. U.S. Department of Agriculture figures show that at the turn of the century, the consumption of what are called "complex carbohydrates"— fruits, vegetables, and whole grains—occupied a significantly larger part of our diet in this country. Since 1910, our consumption of carbohydrates has fallen, while our intake of fats has risen. In the three decades between 1947 and 1976, our per capita potato consumption dropped 21 pounds per year, our flour consumption 31 pounds, our vegetable consumption (excluding green and deep yellow vegetables) 12 pounds, our consumption of green and deep yellow vegetables 6 pounds. Our bread consumption fell in the same period. But in many countries outside the Western world, a diet high in complex carbohydrates such as beans and grains and the foods made from them is still standard fare. Thus a switch to a diet higher in carbohydrates and lower in fat is not really a change but a return to eating patterns that have prevailed throughout most of the world's history, including that of our own nation.

Most of the features of the new diet nutritionists are recommending to combat heart disease and cancer appeared in a diet devised some twenty years ago by the late Dr. Norman Joliffe, who was then the director of the Bureau of Nutrition of the New York City Department of Health. Joliffe's diet, which he named the Prudent Diet, was used as the basis of a fourteen-year study of a large group of middle-aged New York City men. Those who followed the diet had about one-half the rate of heart attacks that would have been expected in a group of similar age. The Prudent Diet is a fairly strict regime, allowing only one pound of red meat (excluding veal) per week. This means that the bulk of the meat on the diet is made up of fish and poultry, a feature many Americans would dislike. Even if you don't follow the diet rigidly, however, it offers many helpful suggestions, including a recipe for a solid shortening based on safflower oil that is low in saturated fats. The principles of the diet, together with a large number of recipes, are incorporated in a 1973 book, *The Prudent Diet* by Iva Bennett and Martha Simon, but the book, alas, is already out of print. I found a copy at my library and you may be able to unearth one there, too.

Someday we may be able to obtain beef and dairy products with less saturated fat or eat high-cholesterol foods and excrete all or most of the cholesterol. Research is progressing on both approaches. At the U.S. Department of Agriculture's Agricultural Research Service headquarters in Beltsville, Maryland, Dr. Joel Bitman is developing a method to reduce saturated fats in beef, veal, cheese, and milk by feeding cattle droplets of unsaturated fat encapsulated in protein. The capsules prevent bacteria in the animals' stomachs from converting the unsaturated fat to saturated fat, with the result that the unsaturated fat passes into the cow's lower intestinal tract and is digested in that form. By means of this process, the ARS researchers have reduced the saturated fat content of milk by 33 percent, of beef by 18 percent, and of veal by 14 percent. Work is underway at the University of Cincinnati on sucrose polyester, or SPE, a compound made from soybean oil and sugar which appears to reduce blood cholesterol levels when fed to human subjects along with high-cholesterol food. A group of fifteen subjects with normal cholesterol levels who ate SPE along with a high-cholesterol diet showed a 16 percent drop in blood cholesterol levels, the SPE and some cholesterol having been excreted by the body. SPE can be substituted for cooking oil or added to high-cholesterol food.

Even if science enables us to reduce the fat in foods of

animal origin, however, a change in our affluent diet seems imminent. "Few potential social policies promise so many benefits and so few costs as the decision to alter the affluent diet," says "Worldwatch Paper 9." The alternative protein sources will play a big role in that change.

SUGGESTED READING

All About Cooking Beans, Michigan Bean Commission, 921 N. Washington, Lansing, Mich. 48906, 1976. Stresses canned beans.

Amaranth Round-Up, Rodale Press, 33 E. Minor St., Emmaus, Pa. 18049, 1977 (free, as long as supply lasts).

Bean Cuisine, Beverly White, Beacon Press, 25 Beacon St., Boston, Mass. 02108, 1977. Vegetarian.

The Book of Tofu, William Shurtleff and Akiko Aoyagi, Autumn Press, 7 Littell Road, Brookline, Mass. 02146, 1975.

The Buffalo Gourd, W. P. Bemis and others, Technical Series Bulletin No. 15, Agency for International Development, Washington, D.C. 20523, 1975.

Butterflies in My Stomach, Ronald L. Taylor, Woodbridge Press, P.O. Box 6189, Santa Barbara, Calif. 93111, 1975.

California Ways with California Dry Beans, Genevieve Callahan, California Dry Bean Advisory Board, P.O. Box 943, Dinuba, Calif. 93618.

The Complete Book of Pasta, Jack Denton Scott, Bantam Books, 666 Fifth Ave., New York, N.Y. 10019, 1968. Many recipes using pasta with alternate sources of protein.

Construction of a Chemical-Microbial Pilot Plant for Production of Single-Cell Protein from Cellulosic Wastes, C. C. Callihan and C. E. Dunlap, Superintendent of Documents, U.S. Government Printing Office, Washington, D.C. 20402, 1971 (Stock No. 5502-0027).

Countryside, 312 Portland Rd., Highway 19 East, Waterloo, Wis. 53594. Magazine for owner of one or more acres who aims for self-sufficiency.

Crawfish Farming, Louisiana Wildlife and Fisheries Commission, P.O. Box 44095, Capitol Sta., Baton Rouge, La. 70804, 1970.

Crops and Man, Jack R. Harlan, American Society of Agronomy, 677 South Segoe Road, Madison, Wis. 53711, 1976.

The Dairy Goat Journal, Box 1908, Scottsdale, Ariz. 85252.

Diet for a Small Planet, Frances Moore Lappé, Ballantine Books, 201 E. 50th St., New York, N.Y. 10022, revised edition, 1975. Equal parts nutrition, vegetarian recipes stressing supplementation.

Domestic Rabbit Cookbook, American Rabbit Breeders Association, Box 426B, Bloomington, Ill. 61701, 1976. Numerous recipes, plus cooking tips.

Domestic Rabbits, American Rabbit Breeders Association, 10007 Morrissey Drive, Bloomington, Ill. 61701. Membership in ARBA includes this bimonthly journal.

Entertaining with Insects, Ronald L. Taylor and Barbara J. Carter, Woodbridge Press, P.O. Box 6189, Santa Barbara, Calif. 93111, 1976.

The Farm Vegetarian Cookbook, The Book Publishing Company, Summertown, Tenn. 38483, 1975. Strict vegetarian cookbook.

Fats, Oils and Cholesterol, Carlson Wade, Pivot Original Health Books, Keats Publishing Company, 212 Elm St., Box 876, New Canaan, Conn. 06840, 1973.

The First American Peanut Growing Book, Kathy Mandry, Random House/Subsistence Press, 201 E. 50th St., New York, N.Y. 10022, 1976. Tips on growing peanuts indoors in pots, some recipes.

Fisheries of the United States, National Marine Fisheries Service, Superintendent of Documents, U.S. Government Printing Office, Washington, D.C. 20402, 1976 (Stock No. 003-020-00138-6).

The Food in Your Future, Keith C. Barrons, Van Nostrand Reinhold Company, 450 W. 33rd St., New York, N.Y. 10001, 1975.

Food of Our Fathers, Institute of Food Technologists, 221 N. LaSalle St., Chicago, Ill. 60601, 1976.

Harrowsmith, Camden House Publishing, Camden East, Ontario, Canada KOK 1 JO. Organic farming.

Homesteader's Handbook to Raising Small Livestock, Jerry Belanger, Rodale Press, 33 E. Minor St., Emmaus, Pa. 18049, 1974.

Human Nutrition, Benjamin T. Burton, McGraw-Hill, 1221 Avenue of Americas, New York, N.Y. 10020, 1976.

Legumes in Human Nutrition, W. R. Aykroyd and Joyce Doughty, Food and Agriculture Organization of the United Nations, 1964 (order from Unipub, Box 433, Murray Hill Station, New York, N.Y. 10016).

Magnificent Microbes, Bernard Dixon, Atheneum, 122 E. 42nd St., New York, N.Y. 10017, 1976.

Making Aquatic Weeds Useful: Some Perspectives for Developing Countries, National Technical Information Service, Springfield, Va. 22161 (NTIS Accession No. PB 265-161), 1976.

The Malnourished Mind, Elie A. Shneour, Doubleday/Anchor, 245 Park Ave., New York, N.Y. 10017, 1975.

Malnutrition, Learning and Behavior, Merrill S. Read with David Felson, National Institute of Child Health and Human Development Office of Research Reporting, National Institutes of Health, Bethesda, Md. 20014, 1976.

Meatfacts, American Meat Institute, P.O. Box 3556, Washington, D.C. 20007, 1976.

Michele Evans' All Poultry Cookbook, Dell, 1 Dag Hammarskjold Plaza, New York, N.Y. 10017, 1974. New and old ways to prepare chicken.

The Mussel Cookbook, Sarah Hurlburt, Harvard University Press, 79 Garden St., Cambridge, Mass. 02138, 1977.

New York Times Natural Foods Cookbook, Jean Hewitt, Avon Books, The Hearst Corp., 959 Eighth Ave., New York, N.Y. 10019, 1972. Comprehensive and nonvegetarian, with emphasis on grains, beans, seeds, roots, organ meats.

Nutritional Improvement of Food Legumes by Breeding, Max Milner, ed., Protein Advisory Group of the United Nations, John Wiley & Sons, 1 Wiley Drive, Somerset, N.J. 08873, 1973.

Oats, Peas, Beans and Barley Cookbook, Edyth Young Cottrell, Woodbridge Press, Box 6189, Santa Barbara, Calif. 93111, 1974. Vegetarian, some nutritional information.

Organic Gardening and Farming, Rodale Press, 33 E. Minor St., Emmaus, Pa. 18049. Monthly, with organic orientation. Often includes material on alternate sources of protein.

The Peanut Cookbook, Dorothy C. Frank, Clarkson N. Potter, distributed by Crown Publishers, 419 Park Ave. S., New York,

N.Y. 10016, 1976. Small, well-chosen selection of recipes, with and without meat.

Protein: How Much Is Enough?, Jean Mayer, M.D., Newspaperbooks, P.O. Box 259, Norwood, N.J. 07648, 1975. Facts about proteins from a famous nutritionist.

Protein Resources and Technology: Status and Research Needs, Nevin S. Scrimshaw, Daniel I. C. Wang, Max Milner, Superintendent of Documents, U.S. Government Printing Office, Washington, D.C. 20402, 1975 (Stock No. 038-000-00251-1).

The Prudent Diet, Iva Bennett and Martha Simon, David White (out of print), 1973.

Raft Cultivation of Mussels in Maine Waters, Richard A. Lutz, Maine Sea Grant Bulletin 4, Ira C. Darling Center, Walpole, Maine 04573.

The Role of Animals in the World Food Situation, Rockefeller Foundation, 1133 Avenue of Americas, New York, N.Y. 10036, 1975.

Scientific American, September, 1976, 415 Madison Ave., New York, N.Y. 10017. Issue on food and agriculture.

Selecting and Raising Rabbits, Superintendent of Documents, U.S. Government Printing Office, Washington, D.C. 20402, 1972 (Stock No. 001-000-02640-2).

Single Cell Proteins from Cellulosic Wastes, W. D. Bellamy, General Electric Company, Corporate Research and Development, P.O. Box 43, Schenectady, N.Y. 12301, 1974.

Southwestern Missouri Dairy Goat Association Cookbook, Rt. 1, Box 545, Springfield, Mo. 65803. Covers everything from meat to ice cream.

The Soybean Cookbook, Dorothea Van Gundy Jones, Arc Books, 219 Park Avenue S., New York, N.Y. 10003, 1970. Vegetarian.

Sunflowers: Production, Pests, and Marketing, Extension Bulletin 25, June, 1975, North Dakota State University, Fargo, N.D. 58102.

That We May Eat, Yearbook of Agriculture, 1975, U.S. Department of Agriculture, Superintendent of Documents, Government Printing Office, Washington, D.C. 20402. Issue on agricultural research.

The Two Faces of Malnutrition, Worldwatch Paper 9, Erik Eckholm and Frank Record, Worldwatch Institute, 1776 Massachusetts Avenue, N.W., Washington, D.C. 20036, 1976.

Underexploited Tropical Plants with Promising Economic Value, National Technical Information Service, Springfield, Va. 22161 (NTIS Accession No. PB 251-656), 1975.

The Winged Bean: A High-Protein Crop for the Tropics, National Technical Information Service, Springfield, Va. 22161 (NTIS Accession No. PB 243-442), 1975.

World Food and Nutrition Study: the Potential Contributions of Research, National Academy of Sciences, Printing and Publishing Office, 2101 Constitution Avenue, N.W., Washington, D.C. 20418, 1977.

World Population Trends: Signs of Hope, Signs of Stress, Worldwatch Paper 8, Lester R. Brown, Worldwatch Institute, 1776 Massachusetts Avenue, N.W., Washington, D.C. 20036, 1976.

MORE HELP

The following organizations offer information on specific alternative sources of protein.

AMERICAN GOAT SOCIETY
1606 Colorado Street
Manhattan, KS 66502

Information on rearing goats; ask for sheet describing publications.

AMERICAN RABBIT
  BREEDERS
  ASSOCIATION
10007 Morrissey Drive
Bloomington, IL 61701

Information on rearing rabbits.

AMOCO FOODS COMPANY
200 East Randolph Drive
Chicago, IL 60601

Brochures on Amoco yeast.

CALIFORNIA DRY BEAN
  ADVISORY BOARD
PO Box 943
Dinuba, CA 93618

Pamphlet on bean nutrition.

COUNCIL FOR AGRICUL-
  TURAL SCIENCE AND
  TECHNOLOGY (CAST)
Agronomy Building
Iowa State University
Ames, IA 50011
515-294-2036

Provides names of agricultural experts, information on various aspects of agriculture.

FARM FOODS
The Farm
156 Drakes Lane
Summertown, TN 38483

Variety of items, including tempeh and tofu kits, soybeans, tempeh starter, TVP, yeast. Write for brochure.

GARDEN WAY PUBLISH-
ING COMPANY
Charlotte, VT 05445

Books on raising cows, rabbits, goats, sheep, pigs, poultry. Write for catalog.

NATIONAL LIVESTOCK
AND MEAT BOARD
36 South Wabash Avenue
Chicago, IL 60603

Catalog describes publications on meat, including charts, visual aids, pamphlets.

NATIONAL MARINE
FISHERIES SERVICE
Marketing Services Division
Dale Avenue
Gloucester, MA 01930

Poster, "Joy of Cooking Squid," with recipes; other recipes for underutilized marine species.

THE POTATO BOARD
1385 South Colorado Blvd.
Suite 512
Denver, CO 80222

Information and recipes for potatoes.

VERMONT BEAN SEED
COMPANY
3 Ways Lane
Manchester Center, VT 05255

Free catalog of beans and peas, many of hard-to-find varieties.

INDEX